Praise for

Mind, Matter and Life

I am very glad to suggest the publication of this very unusual book that delves into the mind-matter problem with unparalleled creativity and originality. It bypasses the conventional discussions on the topic by exploring not only the intricate relationship between the human mind and the brain but also its profound implications for the entire phenomenon of life and our understanding of nature in materialistic terms.

The authors' two cutting-edge "examples" (atmospheric electricity and pions interactions) challenge the prevailing neo-Darwinian approach to the origin of life, inviting readers to critically evaluate the claim that naturalistic explanations alone can account for the emergence and development of living organisms. By doing so, they critically surmise the possibility that our current understanding of life may be incomplete.

One of the most commendable aspects of this book is its resolute commitment to intellectual honesty. The authors fearlessly point out the dangers of scientistic materialism, highlighting how it has inadvertently assumed a quasi-religious status in contemporary discourse. By raising thought-provoking questions, they compel us to reassess our assumptions and open ourselves to alternative perspectives.

The book takes a bold stance that demands attention. It suggests that consciousness or mind could be the fundamental essence preceding matter, offering a compelling framework for scientific investigation. As such, the authors contribute to the advancement of our collective understanding of the mysteries surrounding the mind-matter relationship.

I am sure that this book will become a beacon of inspiration for those seeking intellectual stimulation and a broader perspective. It challenges the status quo, encourages critical thinking, and allures readers to embark on a captivating journey of exploration.

In conclusion, I wholeheartedly endorse this remarkable book. Its unconventional approach, insightful analysis, and thought-provoking arguments make it an invaluable resource for anyone interested in the mind-matter problem, the nature of life, and the broader implications of materialistic perspectives.

Giuseppe Trautteur, Professor Emeritus of Theoretical Physics at the Università degli Studi di Napoli Federico II & scientific consultant for Adelphi publishing house

Mind, Matter and Life

The Mind-Matter Problem within the
Neo-Darwinian Materialist Conception
of Nature

Mind, Matter and Life

The Mind-Matter Problem within the
Neo-Darwinian Materialist Conception
of Nature

Massimiliano Sorrentino

Daniela Panighetti

London, UK
Washington, DC, USA

CollectiveInk

First published by Essentia Books, 2025
Essentia Books is an imprint of Collective Ink Ltd., Unit 11, Shepperton House,
89 Shepperton Road, London, N1 3DF
office@collectiveinkbooks.com
www.collectiveinkbooks.com
www.essentia-books.com

For distributor details and how to order please visit the 'Ordering' section on our website.

Text copyright: Massimiliano Sorrentino and Daniela Panighetti 2024

ISBN: 978 1 80341 613 7
978 1 80341 530 7 (ebook)
Library of Congress Control Number: 2023943498

A CIP catalogue record for this book is available from the British Library.

Design: Lapiz Digital Services

UK: Printed and bound by CPI Group (UK) Ltd, Croydon, CR0 4YY
Printed in North America by CPI GPS partners

We operate a distinctive and ethical publishing philosophy in all areas of our business, from our global network of authors to production and worldwide distribution.

Contents

To Arianna,
always in our hearts

Acknowledgements

We are extremely grateful to Professor Giuseppe Trautteur for sharing his thoughts and insights, like a dear friend. A heartfelt thanks goes to Bernardo Kastrup for the trust placed in us.

We would like to extend our thanks to the monks of Eremiti di Cerreto (Venosa, Potenza, Italy), in particular to Father Cesare, Father Tony and Adele, who hosted us during the writing stages; and thanks also to the monks of the Goleto Abbey (Sant'Angelo dei Lombardi, Avellino, Italy). Thanks should also go to all the friends who have been close to us during this lengthy work. A sincere thank-you goes to Claire Power for the meticulousness and thoroughness with which she reviewed and edited the text.

And finally we are also thankful to Maddalena and Enzo who have greatly supported us throughout this project. Many thanks to Valerio for his extraordinarily cheerful attitude.

Abbreviations and Acronyms

A   Adenine
ADP   Adenosine DiPhosphate
AMP   Adenosine MonoPhosphate
ATP   Adenosine TriPhosphate
BNL   Brookhaven National Laboratory
C   Citosine
CCN   Cloud Condensation Nuclei
CERN   Conseil Européen pour la Recherche Nucléaire
CMB   Cosmic Microwave Background
CMP   Citosine MonoPhosphate
CTP   Citosine TriPhosphate
dAMP   DeoxyAdenosine MonoPhosphate
dCMP   DeoxyCitidine MonoPhosphate
dGMP   DeoxyGuanine MonoPhosphate
dNMPs   DeoxyriboNucleoside MonoPhosphates
dNTPs   DeoxyriboNucleoside TriPhosphates
dTMP   DeoxyThymine MonoPhosphate
e.g.   *exempli gratia* (for example)
e.m.   Electromagnetic
eV   Electronvolt
fig.   Figure
fm   Fentometer
G   Guanine
GMP   Guanine MonoPhosphate
GTP   Guanine TriPhosphate
GZK   Greisen-Zatsepin-Kuzmin limit
Hz   Hertz
i.e.   *id est* (that is)
kV   Kilovolt
LHC   Large Hadron Collider
LSM   Linear Sigma Model

MeV   Megaelectronvolt
MIT   Massachussets Institute of Technology
MV   Megavolt
Na-K   Sodium-Potassium
NDPs   Nucleoside DiPhosphates
NMPs   Nucleoside MonoPhosphates
NTPs   Nucleoside TriPhosphates
Pb   Lead
preTP   Precursor TriPhosphate
QCD   Quantum ChromoDynamics
QED   Quantum ElectroDynamics
QGP   Quark-Gluon Plasma
RNA   RiboNucleic Acid
RREA   Relativistic Runaway Electron Avalanche
SPS   Super Proton Synchrotron
T   Thymine
tRNA   Transport Ribonucleic Acid
U   Uracil
UMP   Uridine MonoPhosphate
UTP   Uridine TriPhosphate
W   Watt

Introduction

The mind-matter problem, also known as the mind-body problem or the mind-brain problem, is typically considered to be an issue that concerns the relation between mind and brain in humans and, at most, in those other animals for which a certain neural activity is supposed to be related to subjective experience of mental contents, such as perceptions and thoughts. However, in the introduction to his famous and much-critiqued book *Mind and Cosmos: Why the Neo-Darwinian Materialist Conception of Nature Is Almost Certainly False* (Nagel 2012), Thomas Nagel writes that he wishes to show that "the mind-body problem is not simply a localized problem that affects the relationship between mind and brain in living animal organisms, but rather it pervades our understanding of the whole cosmos and its story", considering that his work is focused mainly on the biological world, in which living organisms consistently give the appearance of being a product of mind. Although we— like many other authors—remain sceptical of Nagel's idea of a vaguely specified natural finalism, we too believe that the gravity of the mind-body problem is currently underestimated, even though we follow a completely different approach to support the same thesis.

Indeed, the purpose of this work is to show that the mind-body problem does not only concern the relationship between mind and brain in humans and in those other animal organisms for which a certain activity of their brain is supposed to be related to subjective experience, but rather it affects the current conception of the whole phenomenon of life and, in fact, the whole conception of nature in materialist terms. In particular, most of this work is focused on discussing the neo-Darwinian conception of the phenomenon of life, specifically addressing the claim that, since a naturalistic explanation of the emergence

1

and development of living organisms on Earth exists or is in principle available, living organisms are not what they might appear: they are not the product of a designer's mind.

Fundamentally, we argue that such a reasoning cannot ignore the following two points:

- the concept of authorship is deceitful and needs to be carefully deconstructed to be used in a rigorous argument;
- non-living systems have existed in the universe since before the first living systems on Earth and therefore may in principle have a causal power over the emergence of living systems, which, by and large, give the appearance of being a product of mind; given the current state of ignorance about the mind-matter relation, we have no rigorous way of establishing whether or not the activity of certain non-living systems can be related to a subjective experience of mental contents.

Specifically, we focus on the classical scenario within which the origin of life is conceived, pointing out that, in such a typical neo-Darwinian narrative about the origin of life, a certain complex self-organizing system — which we will be discussing extensively — has a decisive causal role in the appearance of the first living organisms on Earth. Furthermore, we examine whether such a system can be conceived as a 'brain' of a natural designer, i.e. a system whose activity is related to subjective experience of authorship, inasmuch as the activity of such a system could be interpreted as giving rise to systems which give the strong impression of design. In other words, we claim that, even though the origin and evolution of life can be explained in purely physico-chemical terms within the neo-Darwinian framework of biology, given the current state of ignorance with respect to the mind-body problem it is not straightforward to conclude that living organisms are not what they might appear: products of mind. Equivalently, although the neo-Darwinian conception constitutes a valid naturalistic explanation of the origin of living organisms, a conclusion regarding the actual nature of the phenomenon of life cannot be formulated without taking into account the mind-matter problem.

Subsequently, we point out that if life were actually a product of mind, it would be likely that we would not be able to fully comprehend it through a purely materialist description of the phenomenon, as typically happens for any product of mind. Thus, we attempt to conceive a semantic plane of understanding of what life might be as a product of mind, in order to support the thesis that life might in fact be a product of a natural designer's mind. In this regard, we question present notions pertaining to the emergence of mind in the history of the universe, suggesting its rise could predate current estimates by 4 billion years.

**

Finally, we revisit the mind-matter problem in the light of what has been proposed and argue that, regardless of when subjective experience is considered to have occurred for the first time, it is not possible to provide a rigorous explanation of subjective experience in terms of the activity of any objectively existing physical system within the objective material world as conceptualized in a materialistic perspective. Therefore, we argue that the mind-matter problem not only has the potential to question the conclusions of the neo-Darwinian framework about the phenomenon of life, but also points to certain intrinsic limitations of the current way in which reality is conceived in terms of the concept of matter.

Since much of this work deals with discussing the nature of the phenomenon of life, we must first of all briefly revisit the current understanding of the phenomenon in physico-chemical terms by dedicatedly focusing on certain aspects which are crucial for supporting our thesis.

Chapter 1

The Cell: A System Which Seems to Be a Product of Mind

A living organism is an open thermodynamic system which keeps itself in an internal disequilibrium condition by dissipating the free energy that it absorbs from its environment.

The main source of energy for life on Earth is constituted by the proton-proton chain nuclear reaction

$$4^1H \rightarrow\ ^4He + 2v + 27MeV$$

responsible for helium nucleosynthesis in the Sun's core, while the main process responsible for the absorption of such energy by living systems on Earth is photosynthesis, roughly described by the reaction

$$E + 6CO_2 + 6H_2O \rightarrow C_6H_{12}O_6 + 6O_2$$

In such a process, a glucose molecule is synthesized starting from inorganic compounds by harvesting about $100eV$ of energy carried by visible solar photons — each carrying about $2.5eV$. In brief, the incoming energy breaks the bonds of stable inorganic compounds (CO_2 and H_2O) and is then stored in higher-energy chemical bonds (glucose $C_6H_{12}O_6$), while waste (O_2) is released into the environment.

The process of cellular respiration is instead common to most photosynthetic and non-photosynthetic organisms, and it can be described, first off, as the reverse reaction to photosynthesis

$$C_6H_{12}O_6 + 6O_2 \rightarrow 6CO_2 + 6H_2O + E$$

In such a process, a fraction of the energy stored in a glucose molecule is released back into the environment; for instance, the total dissipated power of a human organism—due to all its dissipative processes—is about 100W.

A photosynthetic organism lacking access to light long enough to exhaust its internal reserves reaches the condition of thermodynamic equilibrium; in other words, it dies, as happens to an animal that is deprived of food or oxygen for a sufficiently long time. More generally, a living organism unable to exchange matter and energy with its environment tends towards the condition of thermodynamic equilibrium—its death—like any other isolated system. Conversely, if free energy is constantly supplied to the system, such free energy sustains a high level of internal organization compared to the environment; in other words, the organism stays alive—until, of course, eventual long-term ageing processes of an entropic nature inexorably push it towards its death. On the other hand, cellular respiration, like combustion, is an oxide-reduction process—where oxygen oxidizes glucose to CO_2 by reducing itself to H_2O—and it is useful to note that even the flame of burning fuel, which requires a constant supply of fuel and oxygen from the environment, is an open system which dissipates free energy to sustain its disequilibrium organization.

In such disequilibrium systems, the energy flowing through the system sustains its internal disequilibrium organization. Such a mechanism was accurately formalized by Ilya Prigogine (Prigogine and Stengers 1984) into the concept of dissipative structures, i.e. structures that arise in dissipative systems, within the formalism of spontaneous symmetry breaking. A typical example of a dissipative structure is the famous Belousov-Zhabotinsky reaction, and many other dissipative structures are observable in everyday life since they are responsible for pattern formation in clouds, such as the Bénard convection cells and the Kelvin-Helmholtz instabilities (Smyth and Moum 2012).

Destructive phenomena such as tornadoes (Bystrai, Lykov, and Okhotnikov 2011) are dissipative structures, and star formation regions can be modelled as galactic dissipative structures too (Bodifée 1986). As pointed out by Anderson et al. (Anderson and Stein 1987), the mathematics of dissipative structures deriving from spontaneous symmetry breaking has remarkable analogies with that of phase transitions, often characterized by spontaneous symmetry breaking; e.g. Bénard's convection cell formation (fig. 1.1a) and the process of ice crystallization (fig. 1.1b) in liquid water cooling are both characterized by hexagonal honeycomb-like structures formed by spontaneous breaking of rotational and translational symmetry. In a dissipative system, as in a phase transition, a uniform state becomes unstable, forcing the system into a patterned one, leading to the self-organization of the system.

a)   b)

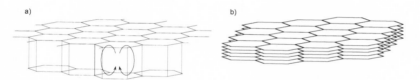

Figure 1.1 a) Bénard's convection cell; b) Ice crystalline structure

Thus, while the universe is inexorably moving towards its thermodynamic equilibrium, the formation of ordered structures in nature occurs as an absolutely common event in different thermodynamic scenarios: in disequilibrium systems as in phase transitions, for example; and in both cases the structuring process is effectively described by the wonderful mechanism of spontaneous symmetry breaking.

It is also well known that the emergence of life on Earth is strictly related to a broken symmetry. The fundamental classes of organic molecules, such as carbohydrates, amino acids and lipids, do not constitute a racemic mixture in organisms, but rather

they are homochiral. Kondepudi and Prigogine (Kondepudi
and Prigogine 2014) also attempted to explain such breaking of
chiral symmetry within the framework of dissipative structures.
Regardless of innumerable attempts to explain the origin of
homochirality, it is widely recognized that homochirality is due
to the stereospecificity of enzymes. Most biochemical reactions
are indeed catalysed by highly specific biochemical catalysts—
enzymes—that typically operate by lowering the activation
energy of a reaction, thus increasing the reaction rate; that's why
most biochemical reaction products are kinetic products rather
than thermodynamic products. The mechanism is therefore the
same that occurs, for example, in a catalytic converter, in which
platinum catalyses the oxidation of exhaust gases, but without
doubt biological catalysts are far more specific than typical
non-biological catalysts. Furthermore, specificity is not the only
peculiar characteristic of enzymatic activity. Most of the enzymatic
activity within a cell is ruled by the binding and release of control
molecules that can in turn be produced by enzymatic activity,
resulting in allosteric regulation mechanisms. Such a scenario
of enzyme stereospecificity and complex allosteric regulation is
certainly peculiar to biochemistry.

To provide an example of the complexity of a biochemical
process, it is useful to discuss the process of cellular respiration
in more detail. It is in fact a complex process which basically
consists of three stages: glycolysis, Krebs cycle and oxidative
phosphorylation, each a complex of reactions regulated
by highly specific catalysts. Of those, the latter—oxidative
phosphorylation—is certainly the most interesting because of
its physics. In this stage, the transport of the electrons in a series
of redox reactions is coupled through molecular proton pumps
to the maintenance of a proton concentration gradient across
the two sides of the inner mitochondrial membrane,[1] so that an
opposite proton flow that tends to equilibrate the concentrations
is established through specific sites on the membrane and a steady

state of concentration imbalance is kept (fig. 1.2). Meanwhile, the ATP-synthase machine—equipped with a stator and a rotor, like a miniaturized turbine about $20nm$ high—works in concert with the aforementioned pumps, coupling the proton motive force to the synthesis of ATP molecules. The rotation induced by a flow of three protons causes conformational changes in the static head of the machines to favour the phosphorylation of molecules of adenosine diphosphate (ADP) to adenosine triphosphate (ATP) by the addition of a phosphate group expelling a water molecule.

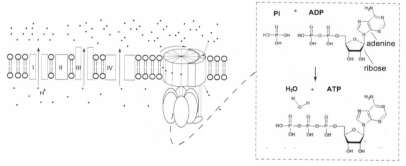

Figure 1.2 Electron transport chain and phosphorylation of ADP catalysed by ATP synthase. ATP is constituted by a nitrogenous base (adenine), a pentose sugar (ribose) and a three-phosphate group

Similarly to cellular respiration, photosynthesis is also divided into stages: the light-dependent phase and the light-independent phase, both occurring in chloroplasts or in photosynthetic bacteria. The light-dependent phase is conceptually very similar to phosphorylation occurring in cellular respiration, with the difference that, in photosynthesis, proton pumping is coupled with the absorption of photons rather than with the oxidation of glucose, so that the ATP synthesis is ultimately coupled to photon absorption and therefore occurs through the photophosphorylation of ADP

$$\gamma + ADP + P_i \rightarrow ATP + H_2O$$

In the light-independent phase, the hydrolysis of ATP

$$ATP + H_2O \rightarrow ADP + P_i + E$$

is coupled with the synthesis of glucose, the fuel that in turn is oxidized during cellular respiration to, once more, produce ATP by oxidative phosphorylation. Basically, ATP constitutes the energy vector for most biochemical processes, since readily available energy is stored into high-energy bonds between its phosphate groups that tend to strongly repel each other because of their negative charge, and the reaction through which such energy is released is the ATP hydrolysis, the energy source for most of the cell's molecular machines. On the other hand, glucose molecules, which can be further packed into glucans, constitute the long-term energy reserve.

Even without delving into every fascinating detail of the processes of photosynthesis and respiration, which are only briefly described here, no one can reasonably deny that such an apparatus looks like an engineering masterpiece, the realization of the 'power plant' of the cell, where the same principles used in our hydroelectric power plants are involved at nanoscale level with protons rather than with huge water flows.

This illustration is just an example of the wonderful world of molecular machines, whose reality exceeds any fantasy, as has often been the case in the history of science. As another example, a machine quite similar to ATP synthase is found on the membrane of flagellate bacteria. In this case, the proton motive force established by respiration is coupled to flagellum rotation, and the rotation of the flagellum causes the motion of the bacterium itself, like a submarine pushed by the rotation of the propeller. Molecular machines can act according to distinct physical principles, behaving like weakly or strongly coupled machines and performing more disparate functions like one-shot machines or cyclic machines, such as motors, pumps and synthases. Moreover, most turbines can work in reverse regime as pumps, just as many human-made machines do. There are molecular motors dedicated to transporting materials and signals on 'railways' constituted by microtubules, while other machines operate to assemble and disassemble these 'railways'; other molecular motors are responsible for muscle contraction in animals, while among the most famous pumps there is the sodium-potassium pump, which is responsible for establishing a voltage gradient of about $-70mV$ in all animal cells between the two sides of the cell membrane by establishing a concentration gradient of sodium and potassium ions, allowing a wide variety of secondary solute transport systems, as well as the action potential discharge in excitable cells.

As we have seen in regard to the processes of photosynthesis and respiration, as well as the others mentioned, the work performed by this machine is crucial for the structuring of the system. There is therefore a dissipative flow of energy in a living organism which sustains its structuring, just as a dissipative flow sustains the structuring of a non-living disequilibrium system; but it is useful to observe that in a living organism a significant fraction of the dissipated energy is dissipated because of the inefficiency of molecular machines, which of course are not present in dissipative structures. One ATP synthase complex can generate more than 100 molecules of ATP each second, and the entire breathing process can 'charge' up to 32 ATP molecules per 'burned' glucose molecule. It is therefore clear that, unlike in a combustion process, the heat released in cellular respiration is in fact the energy that is not harnessed in the ATP molecule, the main energy carrier of the cell, to produce useful work. In absolutely qualitative terms the work done by these machines contributes significantly to the structuring of the system with very high efficiency with respect to the structuring efficiency of a dissipative system. For example, the efficiency of structuring of a tornado or of other dissipative structures is much lower, because the thermodynamic conditions of a tornado are quite different from those of a living organism; typically, dissipative structures appear in far-from-equilibrium systems, while a living organism works much closer to the thermodynamic equilibrium. These general considerations lead to great debates about the relevance of dissipative structures in the problem of the origins of life; regardless of the debate, we can observe that dissipative structures in general do not give the impression of being the result of an intentional project, an impression which is instead very strong with regard to living organisms.

As discussed earlier, molecular machines use the energy released by a certain process (photon absorption, proton flow, ATP hydrolysis) to drive an otherwise unfavourable process (active ion transport, molecules synthesis, various types of mechanical motion) according to different physical principles, and, in fact, most of them couple the hydrolysis of ATP with an otherwise unfavourable process. Since the fundamental characteristics of life are present even in the simplest unicellular organisms, we do not lose any generality in this analysis if we focus on those and, in particular, on unicellular organisms with minimal architecture: prokaryotes. Furthermore, since solar energy is the main source of energy for life—and photosynthesis is the main process through which solar energy is harvested—we will focus on photosynthetic bacteria,[2] such as cyanobacteria.

Below, we will examine two distinct groups of reactions involved in cellular metabolism:

- *anabolic reactions* (e.g. photosynthesis); these are endergonic reactions, i.e. non-spontaneous reactions characterized by an increase in Gibbs free energy ($\Delta G = \Delta H - T\Delta S > 0$) that require free energy (solar photons, ATP hydrolysis) for the synthesis of more structured molecules useful for the growth of the organism;

- *catabolic reactions* (e.g. cellular respiration); these are exergonic reactions, i.e. spontaneous reactions ($\Delta G < 0$) that release energy by breaking down structured molecules into less structured units that can be released into the environment as waste or used as a construction unit in anabolic ways; the released energy can be stored mainly in ATP molecules or other energy carriers and can be used to let anabolic reactions occur.

Typically, in the presence of favourable conditions (light for phototrophic organisms and in general free energy and raw materials), by virtue of the overall anabolism (supported by catabolism) a unicellular organism grows in size, as a praeludium to its reproduction. Thus, in a growing photosynthetic organism the increase in biomass is due to substances acquired from the environment, while the energy that prompts the system of reactions that allow its organization is solar energy. Of course, the above is a fine approximation for practical purposes, but it is not an exact statement. As we said, the free energy of the photons that feed life ultimately comes from the strong interaction field and is released in the p-p chain reaction. Having the Sun with a luminosity of $3.8 \cdot 10^{26} W$, the Sun's mass decreases by $4.2 \cdot 10^9 kg / s$ because of emitted photons; the average rate of energy captured by photosynthesis and stored into chemical bonds is about $10^{14} W$ and the absorbed energy is basically stored into glucose molecules. Therefore, considering that Lavoisier's law is a non-relativistic approximation, the photosynthesis process is responsible for an energy-mass conversion rate of $10^{-3} Kg / s$ on Earth, with which the catabolic process of course concurs. Such a mass excess is mostly contained in biomass since the high-energy molecule produced by photosynthesis (glucose) as well as its metabolic derivatives (ribose 5-phosphate, ATP) are stored in the organism itself. Although the mass excess-defect is absolutely negligible,[3] for practical purposes in biochemistry it is useful to note that metabolism is basically a complex process of energy-mass transformation[4] that in a given reaction

$$\Delta m = \frac{\Delta H}{c^2}$$

can be either positive or negative, being mostly positive for anabolic reactions (responsible for growth), where higher-energy bonds are formed, and negative for catabolic reactions.

Since ATP is obtained by phosphorylation of ADP, during its growth an organism autonomously synthesizes ADP molecules. In fig. 1.3 the complex of reactions responsible for the synthesis of ADP is broadly illustrated: glucose—the hexose sugar synthesized by photosynthesis—is transformed into ribose—a pentose sugar—through an alternative metabolic pathway to glycolysis (pentose-phosphate pathway), while the nitrogenous bases have some amino acids produced by cellular respiration as precursors. Of course, any of these unfavourable processes must be coupled to ATP hydrolysis through a specific ATP-dependent machine in order to occur.

The process shown is just an extract of the whole transformative process taking place within a cell that constitutes cell metabolism. In this process two basic classes of organic compounds are involved, namely carbohydrates (glucose, ribose, etc.) and amino acids (glycine, alanine, etc.), but for the growth of the organism to occur, its membrane—composed essentially of phospholipids—must grow too, and this requires the metabolism of a third class of organic compounds: lipids. The role of the membrane is crucial for the entire cellular metabolism; the whole transformative process is indeed concomitant with the flow of energy and matter through the semipermeable membrane that distinguishes the system from its environment.[5]

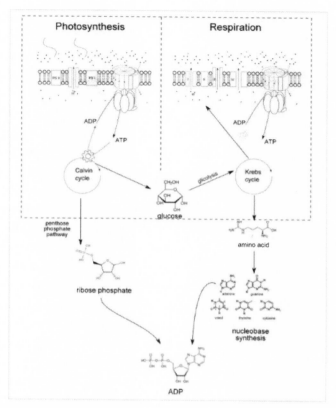

Figure 1.3 Basic metabolic pathways that lead to ADP synthesis

As we have seen, ATP plays a crucial role as energy carrier in metabolism, but this molecule can undergo an alternative pathway: it participates in a polymerization process in which it constitutes a component of a more complex molecule, the RiboNucleic Acid (RNA) molecule, a non-branched linear polymer consisting of a sequence of four distinct riboNucleoside MonoPhosphates (NMPs: AMP, UMP, GMP, CMP) or, in brief, ribonucleotides. The polymerization takes place by virtue of an RNA-polymerase machine which polymerizes an RNA polymer starting from another polymer — the DeoxyriboNucleic

Acid (DNA) polymer—and from free-floating riboNucleoside TriPhosphates (NTPs: ATP, UTP, GTP, CTP). The DNA molecule is a double-stranded antiparallel complementary polymer where each strand consists of deoxyriboNucleoside MonoPhosphates (dNMPs: dAMP, dTMP, dGMP, dCMP), or, in brief, deoxyribonucleotides, which fundamentally differ from RNA monomers due to the different pentose sugar (deoxyribose rather than ribose) and due to the presence of a thymine base rather than uracil. The four dNMPs of DNA are coupled according to the principle of complementarity: a purine base (A, G) is typically bound to the complementary pyrimidine base (T, C). In regard to the free-floating NTPs, we have previously seen how ADP is synthesized and phosphorylated to ATP; the other NDPs are synthesized in similar pathways and phosphorylated to NTPs through catalysts that couple their phosphorylation with ATP hydrolysis. RNA polymerase takes care of opening the DNA molecule by moving forward along a strand, and couples the hydrolysis of two phosphate groups with the polymerization of RNA according to the complementarity principle (fig. 1.4a; in this case an A base is associated with U rather than T). While the RNA molecule is being produced it can undergo the process of RNA folding, in which it folds on itself by virtue of interactions between nucleotides and with water molecules, acquiring a form dependent on the specific sequence of monomers. Therefore, such a form gives the RNA molecule some specific catalytic properties; the catalysts thus obtained constitute a class of enzymes called ribozymes. But alternative pathways are available for RNA molecules (fig. 1.4b): some synthesized RNA molecules (tRNAs) characterized by a specific triplet of nucleotides in a given site are bound to a specific amino acid by the action of aminoacyl-tRNA synthetases (ATP dependent). These machines work in concert with other machines (ribosomes) that instead deal with the reverse process, detaching amino acids from tRNA. In a ribosomal site, an aminoacyl-tRNA molecule binds to a triplet of nucleotides of mRNA according to the complementarity principle. Afterwards, the ribosome advances to the following triplet, allowing the entry of a second aminoacyl-tRNA; then the first amino acid detaches from its tRNA—which is thus released—and attaches to the amino acid of the second aminoacyl-tRNA molecule, which is at this point a tRNA molecule, with a sequence of two amino acids attached. The ribosome then continues to advance along the mRNA, catalysing the synthesis of an amino acid chain releasing tRNA molecules. The ribosomes, thus, synthesize a linear chain of amino acids (a protein) that folds up by virtue of principles similar to those responsible for RNA folding. Finally, the proteins produced can be assembled together by virtue of intermolecular forces to form composite structures such as an ATP-synthase, or to form prominent assemblies of proteins and RNA molecules such as ribosomes themselves.

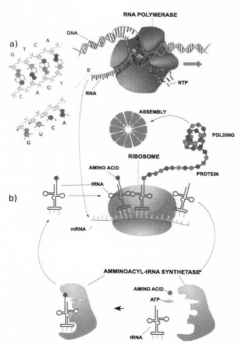

Figure 1.4 a) RNA synthesis; b) Protein synthesis

If, looking at molecular machines, a cell seems to be the result of a project, the impression is even stronger if we look at the way these machines are assembled. Let's suppose we are interested in knowing the amino acid composition of all the proteins that can be synthesized in a given cell. We could certainly analyse the amino acid composition of each protein present in the cell, but it is also possible to proceed in an alternative way. Once the way in which aminoacyl-tRNAs bind specific triplets to a specific amino acid is known, the DNA molecule becomes for us a material medium on which certain triplets of nucleotides represent a given amino acid; i.e. for a subject interested in the cell protein composition, a certain triplet of nucleotides stands for a specific amino acid as a component of a protein (or for the end of the protein). Therefore, a DNA molecule basically constitutes a molecular storage medium for data that contains a certain amount of information which specifies the amino acid composition of all the proteins that can be synthesized in the cell. Such data—sequences of four distinct symbols organized in triplets—can be copied on another data storage medium and undergo data analysis. The data packet in which the information that specifies an RNA molecule or a protein is encoded is called a gene.[6] Such information is called 'genetic information', and the correspondence rules that define the way the genetic information is encoded in data for a subject interested in knowing the cell protein composition constitutes the so-called genetic code: it identifies the correspondence rules between the 64 possible triplets of nucleotides and 20 amino acids (plus the stop codons which encode for the end of the protein) with degeneration. From this perspective, the polymerization process of RNA from a DNA strand appears as a process in which information is copied from one tape to another by dedicated machines, while protein synthesis appears as a process in which proteins are synthesized with a composition specified by genetic information; the synthesis processes of RNA and proteins are therefore called transcription and translation of genetic information.

Referring back to the comparison between a cell and a flame (as representative of dissipative structures), the cell consumes free energy to create and maintain its organization, just as happens in many self-structuring dissipative systems, although less efficiently. But, unlike dissipative systems, living organisms can be usefully described in terms of information: in a living organism free energy prompts an extremely complex system of chemical-physical processes specified by genetic information. This is a convenient description for practical purposes, but not necessary, in principle, for explaining the process of the life of a cell. We explicitly avoid using it in our preliminary description of protein synthesis to underline that it can absolutely be reducible to lower-level physico-chemical processes; a cell, in fact—however complex—is made of nothing but interacting molecules. Nonetheless, we should point out that we also typically describe complex technological human artefacts in terms of information. For example, in a washing machine, free energy is dissipated by inducing physical processes that are specified by the information contained in the washing program, although—as for a living cell—what happens in a washing machine is reducible to the physics of the electric motor, of the heater resistor, and of the solid-state physics as concerns its electronics.

Given the relevance, for a knowing subject, of the DNA molecule as a molecular data storage medium, it is useful to have a look at the process by which the DNA molecule is synthesized, which is quite similar to the RNA polymerization process we have previously described. In the DNA synthesis process, two DNA molecules are each polymerized starting from a mould formed by a single pre-existing DNA strand according to the principle of complementarity of the bases. Each new DNA double strand is formed by a pre-existing strand and a new synthesized strand; in this sense the replication process is semi-conservative. A whole machine complex is responsible for the replication process; e.g. helicase is responsible for unwinding and, consequently, opening the double helix, while the DNA polymerase is involved in the polymerization of the new strand starting from free-floating dNTPs (fig. 1.5a), which are metabolic derivatives of NDPs. Typically, the replication of DNA molecules proceeds along two replication forks that start from one or more common replication origins (fig. 1.5b) and proceed in opposite directions: specific inhibitory mechanisms prevent a new double strand from being further replicated before the entire replication is complete.

Figure 1.5 a) Detail of the DNA replication process; b) Overall view of DNA replication with multiple replication origins (typical of archaea)

Although the replication mechanism is extremely efficient it is not free from errors, which are corrected by virtue of innumerable error-correction mechanisms during and after DNA replication: they are fundamentally based on redundancy, just like human-designed error-correction mechanisms in data flow.

The way in which DNA is organized in a cell strongly depends on the domain or in general on the taxonomic group which it belongs to, but, as we said, we are referring here to organisms with minimal architecture such as prokaryotes. Prokaryotes do not have a well-defined cell nucleus; they contain a long closed DNA molecule and small DNA rings called plasmids (fig. 1.7a), which are independently replicated. But replication is not the only process affecting DNA: peculiar interactions between DNA molecules occur, and they are responsible for the exchange of genetic materials between organisms.[7] For example, if a viral DNA ring is introduced into a bacteria, the proteins encoded in the new ring will be synthesized by the cell ribosomes. Among these proteins, the Cre-recombinase performs a specific function: it catalyses the integration (transposition) of viral DNA into the bacterial DNA on a specific DNA site characterized by a specific sequence of nucleotides (fig. 1.6). As a consequence of such an integration the viral genetic material will be replicated, together with the original bacterial DNA.

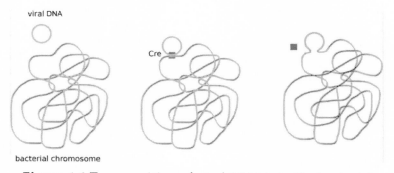

Figure 1.6 Transposition of viral DNA into the bacterial chromosome

We have seen so far the anabolic processes in which the simplest main molecules (glucides, amino acids, lipids) and the most complex ones (nucleic acids and proteins) are synthesized. As we have said, in the presence of favourable conditions (light for photosynthetic organisms and in general free energy and raw materials), a unicellular organism typically grows in size by virtue of overall anabolism, supported of course by catabolism.

However, this growth process cannot continue indefinitely, since this would dramatically increase the size of the organism. There are various factors that limit its size; basically the crucial problem is that the surface-to-volume ratio tends to become unfavourable with respect to the necessity of exchanging matter with the environment. The typical cellular dimensions are therefore determined by the growth rate and the rate of divisions. In fact, the whole growth process is indeed a praeludium to cell division, by which the surface-to-volume ratio is limited. The typical process of cell division for prokaryotes is the binary fission process (fig. 1.7). After DNA replication (fig. 1.7b) is complete, specific mechanisms deal with carrying the two copies of the initial chromosome to the two opposite poles of the cell, while the transmission mechanism of plasmids can be either stochastic (for plasmids present in large numbers of copies) or regulated by appropriate partitioning systems (for plasmids present in double copy or in a low number of copies) which ensure the stable inheritance of plasmids. A contractile protein ring is responsible for the actual scission of the cell. If the whole process is successful then each of the two final systems contains a copy of the whole initial genetic material (fig. 1.7d). As we have seen, DNA determines which proteins are synthesized, and proteins, in turn, determine the transformative processes that take place in the cell; so the two final organisms will essentially be the same as the initial organisms in all those aspects. In other words, the system reproduces itself as faithfully as possible.

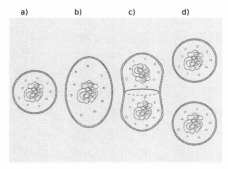

Figure 1.7 a) The parent cell; b) DNA replication; c) Migration of genetic material and formation of the contractile ring; d) The daughter cells

Therefore, among all the processes of synthesis of complex molecules, the DNA replication process, by semi-conservative polymerization, is like the process responsible for the inheritance of genetic information in which the progeny inherits the information of the parent organism.

Each of the two daughter cells will therefore go through the same dynamics of the parent organism, reproducing themselves further, and, in ideal conditions, each organism of the population reproduces itself within a characteristic time frame. Therefore, the population growth rate is proportional to the number of bacteria present at a given time, giving rise to an exponential population growth. Conversely, if a population of bacteria is deprived of its source of free energy, the death phase is observed to be exponential too.

From this brief description of what happens in a unicellular organism, it is clear that the presence of high-technology machines within the systems and the possibility of usefully describing what is going on within the system in terms of information create the strong impression that those organisms are the result of intentional design. Furthermore, a living organism seems to be designed with a technological ingenuity seemingly far more advanced than our current one: the operating principle of machines such as ATP-synthase or a flagellar motor is similar to that of a Feynman-Smoluchowski ratchet, though the latter is powered by a temperature gradient rather than a concentration gradient. But biological nanomachines existed billions of years before humans could even conceive of the possibility of building them. Moreover, even though we are currently able to produce our first synthetic molecular machines—to isolate antibiotics that, for example, block the helicase of bacteria to avoid infections, to synthesize molecules that block the stomach proton pump to avoid stomach acidity, to replace parts of cells with synthetic components, to perform DNA segment replication *in vitro* through polymerase chain reaction, to make useful genetic modifications so that bacteria will produce human insulin for diabetics, or to clone a multicellular organism and so on—we are absolutely far from being able to recreate a living cell from scratch, i.e. to build a living cell starting from water, carbon dioxide, nitrogen compounds and so on.

Such an impression, which is strong even in the simplest, most ancient unicellular organisms, is confirmed if we contemplate the whole tree of life as reconstructed through the integration of fossil and genetic data. More specifically, the impression one has is that some author, who must have been much more skilled and capable than present-day humans, intentionally designed the first forms of life and subsequently intervened with repeated modifications, creating new, more complex ones, suitable for the most disparate environments, more or less similarly to a skilled inventor who creates, for example, the first motor vehicle, and subsequently builds trains with various types of engines, cars, boats, submarines, aeroplanes, jet aircrafts, rockets. In other words, the evolution of living organisms on Earth is a fact, just as the evolution of car models from the 1920s to the present day is a fact: both have changed over time.

The appearance of design is so strong that Richard Dawkins uses it to define what biology is: "Biology is the study of complicated things that give the appearance of having been designed for a purpose" (Dawkins 2017).

But, although this impression is strong, formulating a theory of the origin and evolution of life on Earth by virtue of a designer's intention seems an arduous task for a natural scientist, to say the least. This is because a natural designer is typically thought of as a form of intelligent life—which can in principle also be of extraterrestrial origin—and its introduction only shifts the problem to the origin of this form of intelligent life, which would have intervened in the origin and evolution of life on the Earth about 4 billion years ago. On the other hand, it must be said that this impression has been used, in the history of thought, in support of theological arguments, such as William Paley's famous watchmaker argument, in which he derives the need for an intelligent designer from the evidence of the mechanical complexity of a living organism, just as one might infer the existence of a watchmaker from the complexity of the workings of a clock. However, the introduction of a supernatural designer, specifically a theistic God—as opposed to a deistic God—who intervenes to alter the original plan for the universe, would not give rise to a naturalistic explanation of the emergence of life and would not, therefore, be satisfactory for a natural scientist.

Thus, the current conception of the origin and evolution of life—neo-Darwinism—starts from the assumption that the existence of this designer—the watchmaker of molecular gears—is not necessary in order to explain the origin of life and the differentiation of living species and, consequently, that the appearance of design must be considered only an illusion. In neo-Darwinism it is necessary to postulate that the first entities capable of copying themselves, replicators with certain characteristics, somehow came into being on primordial Earth. According to Dawkins, these replicators must have two fundamental characteristics: occasional errors must occur in the self-replication process, and some properties of the replicators themselves must have an influence on their likelihood of being replicated. Although it is not clear how these replicators originated on primordial Earth, a naturalistic explanation of their advent is in principle available, and, assuming the initial emergence of replicators, nothing else is needed except the mechanism of random mutations and natural selection to give rise not only to the first well-formed cell, but also to all the variety of species existing in the world today. In this regard, Dawkins states that "the theory of evolution by cumulative natural selection is the only theory known to us that is capable, in principle, of explaining the existence of organized complexity". Furthermore, he underlines that the emergence of primordial life-forms and their evolution by natural selection is a blind process, i.e. purposeless and unconscious. In other words, there is nobody who is aware of it. Even if the results give the impression of having been designed, there is no watchmaker except the blind forces of physics governed by the laws of nature, by virtue of which replicators somehow appear and the mechanism of cumulative selection begins. To put it in Jacques Monod's words, living organisms are absolutely not the result of an intentional project, but rather "of chance and necessity" (Monod 1971).

The problem of the origin of living organisms in neo-Darwinism thus seems to be set up in the following way: given that the goal of biology is a scientific explanation of their origin (that is, a theory that is expressed in terms of physics-chemistry), any explanation in terms of intervention by a supernatural entity is not considered a scientifically valid explanation and does not fall within the objectives of biology; on the other hand, the introduction of a natural author of life does nothing but shift the problem backwards. In addition, it is not even necessary to postulate a designer to explain the origin of living organisms, because a scientific explanation for their origin in which an author is not needed is available, or may in principle be available. The neo-Darwinian conclusion is therefore that living organisms are not what they might appear to be, i.e. they are not the product of an author's mind. In other words, the scheme of this deductive reasoning is as follows.

1. Although organized complex systems typically give the strong impression of being the product of an author's mind, if a physico-chemical explanation of their origins is, at least in line of principle, available, then it is not necessary to introduce an author to explain the appearance of design.

2. A physico-chemical explanation of the origins of living systems, which are organized complex systems that give the appearance of design, is, at least in line of principle, available.

3. Consequently, living systems are not the product of an author's mind.

We believe that this reasoning deserves to be investigated further. In particular, it seems appropriate to further examine what we currently know about the nature of the mind of a natural agent in the context of the materialistic paradigm on which contemporary science is based, and also to deconstruct the concept of authorship, which appears to be crucial in the reasoning. So far, however, we have focused on simple single-celled organisms in which the fundamental characteristics of life are found; but to investigate the nature of an author's mind it is necessary to consider more complex organisms in which cellular organization appears in tissues, organs, systems particularly suited for exploring the concept of authorship, and to focus specifically on their most interesting aspect as relates to a possible understanding of the nature of the mind, that is, the nervous tissue.

Chapter 2

The Mind-Matter Problem and the Concept of Authorship

An electrical potential is present across the membrane of all known living cells. Specifically, in all animal cells a membrane potential $v = -70mV$ is established because of a concentration gradient and maintained through the work performed by the ATP-dependent Na-K pump. Furthermore, some of these animal cells — muscle and neural cells — are excitable cells: if subjected to an external stimulus that exceeds a certain threshold, they give rise to a potential impulse, known as the action potential. In muscle cells the action potential triggers muscle contraction, and the external stimulus is typically constituted by the release of neurotransmitters in the chemical synapse between a motor neuron and the muscle cell. Similarly, the impact of a photon on a photoreceptor retinal neuron constitutes a stimulus that triggers the action potential of the neuron; in this case, the resulting action potential constitutes a stimulus that causes another neuron's discharge, whose action potential generally propagates along the neural axon through axonal ionic currents. Like all cellular processes, the action potential is due to the functioning of a complex of molecular devices, which in this case is basically the voltage-dependent channel of sodium and potassium ions (fig. 2.1a).

Without getting into the details of neural electrochemistry, it is, however, clear that what happens in a neuron, in a network of neurons, and in the entire nervous system is reducible to physico-chemical processes; i.e. the activity of the system is reducible to the dissipative energy flow involved in the peculiar electrochemistry of the system and to the functioning of the dedicated molecular devices. If the stimulus continues, the neuron discharges again, giving rise to a train of impulses, which presents a physical quantity—such as the firing rate—dependent on the intensity of the stimulus, thus giving rise to an electrochemical signal that is transmitted along the axon to other excitable cells (fig. 2.1b),[1] including—at the end of the process—muscle cells. The whole process, which starts with a sensory stimulus, such as a visual stimulus, and ends with a motor action, constitutes a sensory-motor circuit. Thanks to these sensory-motor circuits, an animal whose retina is hit by visible photons reacts in some way to the visual stimuli which it is subjected to. For example, a primate—such as a human—can grasp a cup of tea present in its visual field.

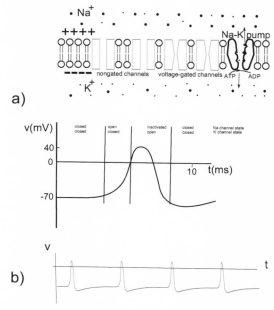

Figure 2.1 a) Action potential generated by neural electrochemistry; b) Pulse train

Let's go back to a subject interested in the study of the brain, as we did in the previous chapter, where we discussed a subject interested in the cell protein composition. For this subject, who observes the entire sensory-motor process, the signals exchanged by the neurons of the living system under examination carry information throughout the system. Therefore, we can also describe the process in terms of information: the brain elaborates information in order to perform a useful task, just as a human-made computing machine endowed with a camera and a mechanical arm, given the initial raw data coming from the camera, computes the function that specifies motion commands in time for grabbing the cup, in synthesis the 'grabbing function'. Even a complex biological system such as a human animal organism therefore allows, as a single cell does, a useful description in terms of information, and consequently gives the appearance of being a product of mind—as do all biological systems, after all. Furthermore, from the study of the brain, as from that of the single cell, it is possible to reconstruct the coding rules of the information carried by neural signals. By observing patterns of activation of the visual cortex, it is nowadays possible to reconstruct, to a certain extent, the visual stimulus which the human organism is subjected to; e.g. a given activity pattern stands for the round shape of the cup seen from the top, while another pattern stands for the rectangular shape of the cup seen from the side. Hence, the whole activity pattern constitutes a nervous representation for an external observer; i.e. for such an external observer, given patterns stand for given stimuli in certain respects, according to reconstructable coding rules. But even in this case the description, in terms of information encoded according to given coding rules, is nothing but a useful description of the process, which is nevertheless ultimately reducible to biochemistry and to the physics of nanomachines, just as the synthesis of proteins encoded in DNA is.

It is important to highlight that, as each of us knows well, when a physical object—such as a cup of tea—is present in the visual field (i.e. visible photons coming from the object hit the retinas, giving rise to the subsequent specific neural activity), each of us experiences our own private perception of the object, unless of course the visual mechanism is somehow compromised. Furthermore, as is well known from neuroscience, there is a close correlation between certain patterns of neural activity observed in a third-person perspective and such private perceptions, the contents of the subject's experience in first-person perspective. In short, when the pattern that encodes the round shape is observed in the visual cortex of our brain (third-person perspective), we actually experience the perception of a round shape (first-person perspective). Moreover, if the cup is too hot we experience a pain sensation, which again strictly corresponds to specific patterns of neural activity, and in this case we typically withdraw our hand from the cup, emitting a cry of pain. Let's therefore suppose we actually build a computing machine which computes the 'grabbing function' according to some generic architecture and we endow it with a camera and a mechanical arm.[2] In addition, the machine is equipped with a temperature sensor with a certain destruction threshold, and if the temperature of the teacup approaches the sensor destruction threshold the mechanical arm suddenly leaves the cup and a loudspeaker emits a cry of pain. The machine imitates with some degree of fidelity a typical human behaviour, and, if we happen to see it in action, the impression that a subject has really got hurt will be very strong. Let's therefore attempt to answer the following question: is there really a subject that has got hurt? Even though very few people believe in an affirmative answer to the question, the point is that whatever definite answer is given to the question—yes, no, very much, a little bit, and so on—there is currently no way to rigorously motivate the answer; therefore the answer—whatever it is—is based on nothing but a personal opinion. The reason why it is not currently possible to rigorously answer the question is that, at present, it is unknown why a certain specific activity of a human brain is related to a specific subjective experience; i.e. it is currently unknown why you actually feel pain when certain specific patterns of neural activity occur in the brain in response to a physical stimulation of your body. Moreover, although the possibility that subjective experience may correspond to the activity of a non-biological material system may seem to someone like a mere fantasy, it is well known today that brain functioning is ruled by the same laws that rule the rest of nature, and a brain is obviously made out of the same stuff which constitutes everything else. It is, then, absolutely hazardous to state that subjective experience can only correspond to the activity of a biological system just because of our ignorance about the principle that underlies this fact. It is therefore not possible to exclude, at least in principle, that subjective experience can be related to the activity of non-biological material systems. Such a possibility has been widely explored in the realm of cognitive sciences and is typically summarized in the famous question "Can machines think?", which investigates whether or not subjective experience of thought can happen in relation to the activity of a computing machine. Thus, if we do not know what makes a certain activity of a physical system—a brain, in this case—correspond to subjective experience, we currently have no rigorous way of establishing—given a certain physical system whose behaviour gives the appearance that there is a subject who is experiencing something—whether or not a certain activity of the system observed in a third-person perspective corresponds to subjective experience in the first-person perspective. In other words, just as we have no way of saying, for example, that "a five-month-old human foetus and an adult dog actually experience pain, while a newborn bat and an old chicken do not", we have no way of rigorously stating whether the cup-grabbing machine experiences some form of pain or not. It is, however, clear that, although we have introduced the issue with respect to the specific experience of pain, our ignorance regarding the topic of experience is absolutely general and independent of the particular content of the experience. For example, referring to the behaviour of the cup-grabbing machine, we equally do not know whether or not 'it' is experiencing the visual perception of the cup, since we absolutely do not know why *we* do. To put it in the words of neuroscientists Kandel et al. (Kandel, Schwartz, and Jessell 1991), the unanswered question we are dealing with is this: "Why does it happen that you see a face when the neurons of the infero-temporal cortex discharge action potentials?"

So far, we have only mentioned three possible contents of experience: pain sensations, visual perceptions, and thoughts. In reality, in the course of everyone's life, a series of contents of experience take place one after the other and overlap, such as — laying out a rough scheme — perceptions (auditory or visual), sensations (touch, taste, smell), emotions, thoughts. In everyday language the thought process is typically referred to with the term 'mind', whereas in the modern philosophy of mind — given that a thought is a content of our experience as well as a visual perception and so on — the mind is typically conceived as the entire content of experience that varies over time, regardless of the fact that we are dreaming or we are awake. In this work we therefore use the term 'mind' in this more general sense. Paraphrasing the metaphorical words of Chalmers, your mind is the whole private movie you experience in your life, every day and every night, enriched with a constant voice-over narrative constituted by the thought process.

In more general terms, the question we are dealing with is therefore: what makes the activity of a brain — which of course is a physical object made of matter — be related to a subjective experience of the mind? Such a question constitutes a formulation of the mind-matter problem, also known as the mind-brain or the mind-body problem.

Admittedly, the answer to this question is currently unknown, even though the most disparate positions have been explored; but for our purposes it is important to figure out what kind of approach to the problem could provide a satisfactory answer. The question we now focus on is therefore the following: is it legitimate to claim that the brain is a computing machine and so to derive an explanation of the subjective experience of mind in terms of computer science? As is well known, this question is linked to the very long yet unresolved debate which followed the criticism posed by John Searle with his famous 'Chinese room' argument against the position he called "strong artificial intelligence". Such a position maintains that a suitably programmed machine that gives the impression of having understood a text actually experiences the understanding of the text.

Of course, to provide an answer to the question, we must focus on what it means to say that a physical machine is a computing machine. Let's consider in this regard the following story, inspired by a famous Zen koan.

Suppose a man takes a universal Turing machine into a forest where there is nobody else. He writes the symbols on the tape to program the machine and the initial data. But the moment he pushes a button to start the machine he dies because of an electric shock caused by a malfunctioning of the system. Regardless of such an accident, the machine is now running fine in the forest, dissipating energy to do work, but there is nobody who is aware of its functioning. Does the machine compute? Our answer is negative, since there is nobody for whom distinct physical patterns on the tape are symbols, and therefore no symbol manipulation occurs. The point is that, in the concept of symbolic representation, representation presumes an experiencing subject, someone for whom something stands for something else; therefore, in the absence of a subject, there is no symbolic representation and no manipulation of symbols. The machine works in the forest like any other physical machine in the forest—like the molecular machines of forest plants—but there is no one in the forest for which that specific Turing machine is a computing machine.

At this point, it is also useful to ask the inverse question: are there machines that for a knowing subject are not computing machines? Suppose that, once the death of the subject in question has occurred, a woman arrives in the forest to contemplate the dramatic scene and she recognizes the machine in operation near the dead human as a Turing machine. It is for her a computing machine, but every other machine in the forest is too, for example ATP synthase: it computes at least the division by three of the number of protons that pass through it, since the number of discrete rotations is the result of dividing the number of protons crossing the membrane by three. Therefore it can be used as a computing machine to divide any number by three, by placing it in an environment with an equal number of protons of which we intend to calculate the division and by counting the number of rotations to obtain the desired result. It is thus easy to verify that every machine computes (in the subject-relative sense in which the term 'compute' makes sense) all the functions with which we can describe the behaviour of the system at some description level and, therefore, that in this perspective a universal computing machine is a machine whose expected behaviour in specific steps can be previously modified by a subject in order to compute any function, where the modification of the expected behaviour is called programming. The behaviour of a physical machine is understood when a mathematical model of the behaviour of the machine is available at some descriptive level. For example, an analogue adder in electronics is an adder in the sense that we can make a physical prediction of the value of the output voltage given by the sum of the input voltages, but in computer science it is a computer that implements the sum function, and this duality of viewpoints is possible for any physical machine. As specified by Searle (Searle 1992), therefore, machines are computers in a subject-relative sense, and in such a subject-relative sense every machine is a computer. Similar considerations apply to the concept of information. Nothing is stored in a flip-flop except electrons and holes, and nothing is stored in a DNA molecule except chemical bond energy, but for a subject for whom each of the two states of the flip-flop stands for the occurrence of each of two equiprobable events, a bit of information is stored in the flip-flop. Similarly, if, for a subject that knows biochemistry, a certain triplet of nitrogenous bases stands for an amino acid, or represents an amino acid, then DNA is a molecular data-storage medium in which a certain amount of genetic information is stored. Computation, information, symbols and, more generally, representation are therefore subject-relative concepts, since a representation is a representation for someone—someone for whom something stands for something else.

It is useful, then, to focus on the term 'machine'. In our mechanistic conception of the world, every physical system is not less machine-like than every machine we can build, and it is a computing machine in the same observer-relative sense. In an example beloved by Searle, a body in free fall is an analogue computer that computes the quadratic function; that is, it can be used to calculate the area of a square whose side is equal to the fall time short of a dimensional constant. It is a computer in the same subject-relative sense in which a personal computer (PC) is a computer, with the difference that a PC is a universal digital computer intentionally designed, while the most interesting function that a free-falling body computes as an analogue computer is the quadratic function and it is not typically conceived as a purposeful designed system.

Therefore, it is not at all difficult to find computing machines in nature, given that basically every physical system is a computing machine in the same observer-relative sense in which our PCs are computing machines: every physical system computes (in the observer-relative sense in which the term 'compute' only makes sense) all the functions with which we can describe the behaviour of the system at some description level. In this perspective, the brain is therefore a computing machine in the same observer-relative sense in which every physical system is a computing machine. But while the bodies in free fall are not surprising at all, our amazement with respect to the study of the brain is comparable to that of a man of the Middle Ages who finds himself dealing with a PC of our times, due to the organized complexity of the system.

Thus, as concerns a possible approach to the mind-brain problem, our position is that the brain is a computer in the same subject-relative sense in which every physical system is a computer. But what about the second part of the question: is it possible to explain the experience of mind in terms of computer science applied to the brain? In other words, is it possible, in line of principle, to formulate a computational theory of mind, such as those engineered by Fodor, Putnam and Lewis, which were attacked by Searle's arguments? Searle's main argument is based on the claim that semantics is not reducible to syntax, but we prefer the following more radical argument. If machines are computers in a subject-relative sense, every explanation of the experience of mind formulated in terms of computer science applied to the brain as a computing machine will suffer from this issue: it will be an explanation of subjective experience which presumes subjective experience itself, without which computation doesn't make sense. It is therefore not possible to rigorously explain subjective experience of mental contents in terms of computer science, which presumes it.

Considerations of this type, which will probably continue to be discussed for many years, draw an epistemological line with respect to the type of explanation that is expected to be found for the experience of the mind, and allow further considerations to be made regarding the problem. If it were possible to formulate a computational theory of mind, this would incorporate the principle of multiple feasibility. A flip-flop can be achieved in many ways, e.g. with a system of hydraulic valves as well as thermionic valves or solid-state transistors, or with a flow of people, and certainly the same computer program can be implemented on computing machines built with very different architectures and based on very different physical principles. If the computational approach to the mind-matter problem were valid, then the architecture and physics of the system would be irrelevant for the subjective experience of the mind to happen. But if, on the other hand, we do not accept any computational theory of mind, subjective experience should be explained only in terms of the objective aspects of the brain — or of another 'thinking machine' — considered real within materialistic ontology. In practice, it should be explained in terms of the physics and architecture of the machine, while taking into account that it is not necessarily true that the architecture of Turing's machine allows the 'miracle' of subjective experience to unfold, especially since the brain presents an architecture that is profoundly different compared to a Turing machine.

It therefore seems that an explanation of the occurrence of subjective experience cannot come from computer science and, consequently, that subjective experience of mental contents must somehow be explained in terms of some biological phenomenon that occurs objectively in the brain. This cannot lead us to believe that the experience of mind must be a prerogative of the brain. As similarly observed by Searle, it is hard to sustain that subjective experience must belong to biological systems alone, because, given that they are absolutely reducible to physico-chemical science, if a biological phenomenon were responsible for subjective experience, it would ultimately be a phenomenon of electromagnetic nature, and as the fundamental interaction involved in all chemical, biochemical and biological processes is the electromagnetic interaction, we can't altogether exclude the suggestion that such a physical phenomenon of an electromagnetic nature could mysteriously give rise to subjective experience of mind in non-biological systems. Searle's thesis is that the experience of mind is caused by brain activity in some way that is not yet understood by neuroscientists, and, therefore, for a machine to have subjective experience, it must duplicate the causal power of a brain to give rise to subjective experience. As a computational theory of mind has been ruled out by previous considerations, we will temporarily take such an approach to the problem for granted and discuss its validity in the final chapter of this work. Let's, however, take a moment to point out that, even though it is not at all trivial to exclude the possibility that, in reference to the cup-grabbing machine, 'it' is experiencing something because it is not biological, we don't know enough to reasonably assert it, since we don't know the nature of such a causal power of the brain from which subjective experience of mind arises. However, when we introduced the cup-grabbing machine, we specified that it is built according to some generic architecture, implicitly assuming that the physics involved is that of the solid state. Certainly, in the approach proposed by Searle, unlike in a computational theory of mind, physics and system architecture are in principle relevant to whether or not there is experience of mind, given that a certain physics and a certain architecture could duplicate the causal power of the brain and therefore cause the experience of mind, while a system that computes the same grabbing function on the basis of a different physics and architecture might not duplicate the causal power of the brain. It is useful to observe, in this regard, that there certainly are different brains that cause subjective experience. Each of us is certain of our own subjective experience and admits that at least other adult humans have subjective experience too; that is, there are some physical machines that are different, to a certain extent, from my brain, but that cause subjective experience too. It is therefore possible, in principle, to make machines that differ from our brain, but cause subjective experience too. What we don't know is to what extent a machine can differ from our brains for subjective experience to occur; e.g. we do not know if the jellyfish nerve net causes some form of subjective experience. Applying this line of reasoning to the case of the cup-grabbing machine, without drawing any definite conclusions, what seems relevant is the physical specifications and the architecture with which the machine is made, rather than which functions the machine computes. It thus seems more plausible that a cup-grabbing machine which is built according to physical properties and with an architecture that share some similarities with those of a brain might actually cause subjective experience, compared to one that, despite computing the same grabbing function, is based on hydraulic valves, as for example a Turing machine.

Obviously, the previous observation is extraordinarily vague because we are in no way able to better specify what is meant by 'similarities in terms of physical properties and architecture', but at the moment we cannot do better, because even admitting Searle's causal approach, we don't know the nature of that causal power. This is at least useful to give an idea of the current state of ignorance regarding a solution to the problem, and this ignorance plays a crucial role in our main thesis, as will become clear shortly.

Now that we have briefly seen what we know and, above all, what we do not know about mind in relation to matter, we need to deconstruct the concept of authorship. So let's suppose that we grab our now familiar cup of tea, burning ourselves and releasing it immediately. Our first-person impression is typically that the sensation of pain is the cause of the sudden action that follows. Generally speaking, our feeling is that our mind has a causal power over the material world that we have conceptualized. In this specific case the impression is that the sudden movement is caused by a mental content, i.e. pain, and, in addition, that we can induce voluntary movements of our body which, in turn, give rise to displacements of other material objects. The crucial point here is that, although we don't understand why we have experience of mental contents, the behaviour of a biological system such as a human organism, as complex as it is, is absolutely reducible in principle to the physics of the system. Therefore, even though it seems to us that the retraction of the hand is caused by the experience of pain, in a careful observation in the third person this experience is not only not observable given the privacy of the mind, but also there is no evidence of unexplained observed events in the causal chain of events that lead to the movement of the arm and that possibly require the invocation of a hypothetical metaphysical mental substance, which would, by the way, trivially make materialism collapse as an ontological paradigm. In other words, there is no well-documented psychokinetic effect detectable, either on physical objects external to our body, or on parts of our body.

In this regard, this aspect of the problem is inextricably connected with the possibility of explaining the emergence of the experience of mind at some point in the evolutionary history of living organisms. What can certainly be explained in terms of evolutionary advantage is the development of sensory-motor circuits whose apparent function is to help preserve the integrity of the organism. However, since the experience of pain is not observable in the third person and, as we have seen, it also does not intervene in the causal chain of events that favour the preservation of the organism's integrity, it is certainly difficult to explain the advent of experience itself in terms of evolutionary advantage, as proposed for example by Crick and Koch (Crick and Koch 1998). For this reason, too, it is hard to rigorously claim that subjective experience can only belong to the biological world, because at the same time it is not at all trivial to look for an explanation of it in neo-Darwinian terms.

Anyway, the question about the potential causal power of mind over matter does not exclusively concern the experience of pain and subsequent movement. For example, when we intend to grab the cup through the movement of an arm, even if the arm actually moves, not only is our intention to move it unobservable in third-person perspective, but also it is not required at any point in the causal chain of physical events that give rise to the movement of the arm and consequently of the cup. What determines the movement or stillness of the cup for a natural scientist can be nothing other than the laws of nature that rule physical events; in other words, physical laws seem to constitute a closed causal system. This is often expressed in the statement: "Mind doesn't have causal power over matter," which can be used as a synthetic expression of a concept; but attention must be paid to this expression in materialism because from a materialistic perspective the ultimate nature of mind must be material in some way or another, and, of course, matter has causal power over matter itself.

To sum up, however much we are convinced that we are the author of the action of moving our arm to grasp the cup, or of moving it whenever we like, in the context of the conceptualization of the physical world in terms of matter, this movement cannot take place by virtue of anything other than the laws of nature that rule the fundamental forces of physics. And what has been said obviously applies to any action, even those much more complex than moving an arm, such as quitting smoking, whose description in terms of willpower — from this point of view — pertains only to folk psychology. It even applies to the creative act involved in building the cup-grabbing machine, of which we would consider ourselves the authors: the machine would materialize by virtue of nothing other than the laws of nature. Clearly, this introduces a huge theme, that of free will. If the system of physical laws is a closed causal system, our free will to build the cup-grabbing machine has no actual causal power over matter, and, therefore, all our constructive choices are determined by an underlying physical dynamic over which we have no power; and this will actually apply to any choice we apparently make in our life. Accepting such a position—the hard deterministic position—for a human being is nothing short of difficult, as emphasized by Giuseppe Trautteur (Trautteur 2009; 2020), because it fundamentally implies the absence of free will, something which Libet (Libet et al. 1983) struggled to accept. On the other hand, the opposite libertarian position, as intuitive as it is and in accordance with our common experience, is much harder to support for a natural scientist.[3] Again, however counterintuitive it may be, according to a rigorous naturalistic analysis it seems that we are justified in believing that we attend to—or witness—any mental content as nothing but bystanders: everything—including what strictly concerns our body and mind—takes place by itself, neither more nor less than the flow of rivers into the sea and the alternation of the seasons caused by the law of gravity, or the ageing of our organism caused by the law of entropy. Nevertheless, unless we are engaged in a careful inner investigation, we tend to consider ourselves the authors of actions that pertain to our body and mind, while of course the same laws of nature responsible for physical phenomena are ultimately responsible for the production of scientific papers on those phenomena, since scientific papers are physical objects. Even though we typically have the impression of being the author of our thoughts and actions, in a third-person description of the process of production of scientific literature, not only is our scientific thought-process not directly observable, i.e. undetectable, but there is also no evidence of unexplainable physical events in the causal chain that leads to the production of the paper, which would require the introduction of a designer. By the way, such a designer's mind would have the same metaphysical location as the mind of a theistic God—however devoid of the superpowers that are typically attributed to it — and every action of such a designer would basically be a small miracle, unexplainable in physical terms. In other words, any designer that intentionally acts upon matter is metaphysical, or supernatural, and we don't need to introduce it, since the system of laws of nature seems — till now — to be a closed causal system, regardless of the fact that such laws rule the occurrence of a single event, rather than the shape of a statistical distribution of them.

Certainly, reconciling the materialistic description of nature, which our body belongs to, with our subjective experience is nothing short of mind blowing, and we have just mentioned the difficulties encountered. It is further useful to consider that most philosophical-scientific works begin with the famous phrase: "The purpose of this work is to...", which expresses the cause of the presence of, for example, a scientific article in finalistic, or teleological, terms. Although such a sentence is quietly accepted by the editors of every adequately peer-reviewed journal, it must be underlined that, in a third-person description of the process of scientific literature production, papers seem to come into being by virtue of nothing but a complex electrochemical process, and the laws of electromagnetism — like any other laws of nature — are, by assumption, absolutely purposeless, or blind, using a word preferred by Dawkins, i.e. they have no apparent purpose, like a blind watchmaker.

Having clarified these problematic aspects, it is useful to ask the question: what does it mean when we say that something, such as a scientific work, is the product of mind? We can define 'products of mind' as those objects whose origin has a full and complete explanation based on a sequence of physical events, even though a certain part of the physical process of production — typically a certain neural activity — is related to subjective experience of intentionality, teleological planning, authorship and, typically, of free will by a subject, who by definition is identified as the author of the work.

In this regard it is helpful to note that, even though at the moment no one is able to produce life from scratch, i.e. to produce a living organism from water, carbon dioxide and so on, there are different forms of life that do have a designer. In fact, humans have been producing genetic modifications for decades, starting with the first rough atomic gardening techniques — in which random mutations in plants were induced by ionizing radiation — and continuing with the most refined modern genetic engineering techniques. Probably, within a few decades, it will be reasonably possible to realize a living organism from scratch and there will be a human designer of a form of life made from scratch; of course, that organism too will be the product of a mind. These organisms produced by humans are, and will be, in fact, the product of mind; that is, they exist by virtue of nothing but the laws of nature, even though the third-person description of the physical events that lead to these forms of life is accompanied, correlated, with a first-person description, in terms of intentionality and planning. It follows that in such cases the two explanations for the existence of a form of life coexist and constitute nothing more than the third-person and first-person descriptions of the same process.

Having briefly discussed what we know about mind in materialism and what it means that someone is the designer of a product of their mind, we believe that we have gathered sufficient elements to resume — in the next chapter — the reasoning with which we concluded the previous chapter regarding the neo-Darwinian assertion on the nature of the phenomenon of life.

Chapter 3

The Mind-Matter Problem
and the Origin of Life

On the basis of the elements collected in the previous chapter, let's comment on the three points behind the reasoning explained at the end of Chapter 1, concerning the neo-Darwinian conclusion about the nature of the phenomenon of life.

1. For a natural scientist, a physico-chemical explanation of the emergence of any physical system is always in principle available, whether it is the development of an element typically considered natural, such as a mountain, or one considered artificial, such as our cup-grabbing machine. Therefore, it is never necessary to introduce a designer to explain the existence of any object or physical system.

2. Therefore, a physico-chemical explanation of the emergence of living organisms must exist, including of course those produced by human authors, regardless of the methodology used: induced random mutation, genome editing, and so on.

3. On the basis of natural science it is therefore not possible to conclude that (non-human-made) living organisms are absolutely not the product of a designer's mind just because a designer's mind—whether it is there or not—is not involved in any naturalistic explanation of the origin of anything physical. This means that—however counterintuitive this may appear—there is never any necessity to introduce mind to explain anything in terms of natural science. Consequently, the existence of a naturalistic explanation for the origin of something that

strongly seems to be a product of design can never be used to argue that in fact it is not.

There is, then, a flaw in the typical reasoning that excludes the possibility that living organisms are the products of mind, and the flaw is due to the fact that the concept of authorship can be misleading if not carefully analysed. Anyway, a natural author's mind is typically thought of as the mind of some form of intelligent life; so, introducing a designer in order to explain the origin of (non-human-made) living organisms would simply yield a regress of the origins of this author. Nevertheless, such an author does not seem to find any place within the neo-Darwinian framework within which the origin and the evolution of life are conceived. Therefore, the more plausible conclusion about the origin of living organisms still seems to be that living organisms are not what they seem to be: they are not the products of design.

We want to scrutinize whether it is actually so easy to reach such a conclusion, and for this purpose we will consider a well-known story that tries to shed light into the 'black box' within which life on Earth would have originated. With respect to the origin of life we use the term 'story', since—even though the theory of cumulative selection is considered by most a scientific theory—it is generally accepted that nothing but a plausible story can be presented about the origins of life. As Dawkins points out, it is clear that it is only possible to speculate about the origin of cumulative selection, because it happened 4 billion years ago in a completely different environment from the current one, and therefore we cannot know anything about the origins of life with any degree of certainty. However, there are a myriad possible speculations on the topic, and none of them requires the introduction of a designer. But since the explanation for the existence of this book also does not ultimately require the introduction of one or more authors—even though we guarantee that in fact we are its authors — we are interested in investigating whether there is at least one story of the origin of life on Earth in which it is possible to identify an author, although their mind, like ours, has no causal power over matter in the sense we specified in the previous chapter.

Modern abiogenetic stories are more refined versions of the old idea of 'spontaneous generation', according to which life was formed from inorganic matter by virtue of the action of mysterious vitalistic forces, which obviously cannot be introduced in modern stories. These can be divided into two large families, the first being the family of stories derived from Oparin-Haldane's work, based on a primordial broth constituted by primordial Earth's ocean. The second family is made up of non-Oparinian stories, which require specific local conditions, such as Marcello Barbieri's ribotype theory, Larry L. Hench's bioactive substrate theory, G. C. Smith's clay hypothesis, G. Wächtershäuser's iron–sulphur world theory, which suggests life evolved inside hydrothermal vents, or C. R. Woese's theory, according to which it is far more likely that life originated from a prebiotic medium that is intrinsically 'cellular' rather than from the uniform ocean of Oparin.

Every Oparinian story postulates the existence of a primordial ocean, as well as the existence of a primitive atmosphere with a strong reducing character, composed of hydrogen, water vapour, methane, nitrogen and ammonia. In such an atmosphere the solar ultraviolet radiation and the electrical discharges of lightning would have caused the synthesis of organic compounds, such as amino acids, purines and pyrimidines. These compounds, dispersing in the oceans, would have formed the so-called 'prebiotic broth', in which the first biomolecules and finally the first living organisms would have formed through a sequence of chemical reactions. The famous Miller-Urey experiment began an experimental tradition that confirmed, to some extent, Oparin's hypothesis. In these experiments, gaseous mixtures similar to the one proposed by Oparin were subjected to the action of electric discharges that simulated atmospheric discharges on primordial Earth, giving rise—as reaction products—to amino acids and other prebiotic precursors. The results suggested that, if the experiment were conducted in a laboratory as large as Earth for millions of years, these precursors could actually organize themselves into self-replicating molecules or into something else that we recognize as living. Although it is legitimate today to doubt the scenario proposed by Oparin for various reasons, it is nevertheless widely recognized that lightning discharge must have been responsible at least for the prebiotic fixation of nitrogen—necessary for the birth of the first organisms—which occurred by virtue of atmospheric and volcanic lightning on primordial Earth (Franzblau and Popp 1989; Raymond et al. 2004). Still today, 10% of the nitrogen fixed in nature that enters the nitrogen cycle is due to atmospheric discharges.

It is therefore unquestionable that, within the little-known causal chain that gave rise to the first life-forms, atmospheric discharges figure as a necessary link in that chain. Even among the family of non-Oparinian stories there is at least one story in which the atmospheric system appears to have a decisive role in the rise of life on Earth. In particular, in the scenario proposed by C. R. Woese—who, by the way, was the discoverer of archaea, and of the horizontal gene transfer, as well as the author of the RNA world hypothesis, and who did not believe in the existence of an ocean on primordial Earth—the atmospheric system plays a much more crucial role than that of a mere electrostatic generator whose activity affects prebiotic chemistry. As we mentioned, Woese believed that it is far more likely that life originated from a prebiotic medium, which is intrinsically 'cellular', rather than from the uniform ocean of Oparin, and that atmospheric clouds would be the best candidate as they are made up of water droplets whose size is typically in the order of tens of micrometres, thus comparable with the size of living cells. Woese's idea was subsequently taken up again by Dobson as an in-depth analysis of the scenario proposed by Oparin, which assumes the existence of the ocean, and they too estimated that, for various reasons, droplets in the range of $0.1–5\mu m$ would have played a much more crucial role than bigger droplets.

Regardless of the presence of the ocean and of the fine size of such cellular objects, let's just consider the different behaviour of molecules interacting with the water molecules of cloud droplets. The hydrophilic molecules present in the atmosphere tend to disperse within the droplets, while the hydrophobic molecules tend to remain outside or to fold on themselves inside the droplet; the amphiphilic molecules — generally equipped with a hydrophilic head and a long hydrophobic tail — tend to arrange themselves by constituting a single molecular layer on the air-water interface (fig. 3.1a), thus generating an inverted micelle structure. In the contact regions between single-layer coated droplets, however, the formation of double-layer regions occurs, and therefore, by virtue of the properties of such amphiphilic double layers, droplets entirely coated by an amphiphilic double-layer membrane can form (fig. 3.1b). These water droplets are among the main constituents of clouds that are electrified in thunderstorms by a charge separation mechanism. Therefore, these droplets are charged droplets at least in certain cloud regions, and, if they are negatively charged, as typically occurs in low-cloud regions where the droplets are more abundant, they tend to attract atmospheric positive ions. This causes an electric potential to be present across the membrane of such droplets (fig. 3.1c), as it is present across the membrane of every living cell. It would be interesting to investigate whether the resting membrane potential of living cells — of an electrostatic nature — can be considered to have somehow evolved out of such potential of an electrostatic nature. Anyway, by virtue of such processes, the first systems appeared on Earth, whose physical size, chemical composition and electrical polarization are comparable with that of living cells.

It is useful to observe that the forces acting on these microsystems cannot be explained by gravity alone, nor by winds, but the action of electrophoretic forces on charged droplets and the action of dielectrophoretic forces even on neutral droplets must be considered, because of the presence of generally non-uniform electric fields within clouds, between clouds, and between a cloud and the ground (fig. 3.1c). Dielectrophoretic forces depend on the size of the droplet and its dielectric constant relative to the air, and in turn the dielectric constant depends on the composition of the solutes. The dimensions of these droplets are precisely those for which the dielectrophoretic forces become important, where for smaller dimensions the Brownian motion dominates and for larger dimensions gravity dominates; the clouds' electric fields reach values of the order of $10-100kV/m$. These are consistent with the electric field values involved in everyday dielectrophoresis techniques at submillimetre scale, which are useful in many fields because of their high selective power. Therefore, these forces can exert a selective action on these microsystems with respect to the overall electric charge, the size and the chemical composition, similarly to how they exert it in human-made electrophoresis or dielectrophoresis.

Furthermore, by virtue of the action of the forces acting on them, these structures can split and merge, giving rise to chemical diversity (fig. 3.1d). And, because of the hydrophilic nature of the surface of these objects, water vapour can further condense on them; thus they naturally become immersed in an aqueous environment (fig. 3.1e). In other words, such coated aqueous objects further act as cloud condensation nuclei (CCN). It is, by the way, well known that still today bacteria can act as cloud condensation nuclei as well as ice nuclei (IN), promoting the formation of ice crystals in the atmosphere (Bauer et al. 2003).

In such conditions ultraviolet radiation constituted by photons whose energy is about 7eV would induce the first metabolic pathways. Depending on whether or not there is an ocean below, these first objects can fall with the rain into the ocean and subsequently be brought back into the atmosphere by the action of winds on the ocean surface and by upward currents.

In such a scenario, therefore, the first organisms based on some primordial form of replicator would appear within the atmospheric water cycle. Such an event would trigger the mechanism of random mutation and natural selection, giving rise in successive steps to an organism more or less similar to a current prokaryote, based on DNA and protein. In the context of random mutations, it should not be overlooked that at high altitudes the radiation from cosmic rays is more intense, and that lightning discharges also constitute a natural source of ionizing radiation, generated by terrestrial gamma ray flash (TGRF).

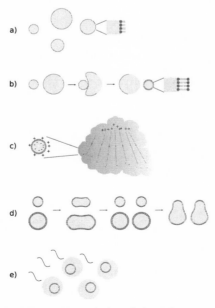

Figure 3.1 Dynamics of cloud droplets coated with surfactant films

Provided, therefore, that in such a scenario a well-formed primordial organism—more or less similar to a prokaryote—may actually appear at a certain point within the tropospheric water cycle, although the organism gives the strong impression of being a product of mind, it is not absolutely necessary to introduce the mind of a designer to explain its emergence, because this occurs by virtue of the purposeless forces of nature. Nevertheless, the question we are interested in is the following: once we believe this is a plausible story of the origin of life on Earth, can we strictly exclude the possibility that such a well-formed primordial form of life that comes into being at a certain point of the process—and that gives the strong impression of being a product of mind—is in fact the product of an author's mind? At first glance, the answer would be affirmative, since nobody we would typically recognize as an author is ostensibly present in the whole scenario, and therefore it seems that there is no place for it; but in fact, to strictly exclude that such a primordial form of life is a product of mind, we should rigorously exclude that—given the crucial role that Earth's atmospheric system plays in the process—there is subjective experience of authorship related to its activity, just as there is such an experience in correspondence with the neural activity of a human author, who equally plays a crucial role in the origin of a product of their mind.

If, indeed, there were experience of authorship related to the system of atmospheric discharges that over the course of millions of years produced the first well-formed living organism, the whole scenario of the emergence of life on Earth would be analogous to that of a chemist who has been playing around for years in his laboratory trying to produce a living organism. Although in this case too the emergence of the living organism occurs by virtue of purposeless electrochemical discharges in the human brain, we know that the produced organism would be the product of the chemist's mind. Can we therefore consider it implausible that there is experience of mental contents related to the activity of the atmospheric system, and archive the question? Regarding this point, it is worth noting that, although it is considered an absolutely legitimate question to ask whether the subjective experience of mind can occur in correspondence with the activity of a non-biological computing machine, the question typically refers to human artefacts and not to natural systems. In this regard it is useful to consider once more that non-biological natural systems are computing machines in the same subject-relative sense in which a human-made Turing machine is a universal digital computing machine, and therefore the question "Can machines think?" can equally refer to the atmospheric system, which, from now on, we will indifferently refer to as 'the atmospheric machine'. By the way, let us note that according to the current neo-Darwinian conception of the biological world, as much as an organism endowed with a brain seems to be designed by a mind, it is not in fact considered to be an artefact, and we know that at least for human brains subjective experience certainly occurs, even though humans can't yet explain it. Therefore, given that the human brain— the only physical system whose activity is certainly related to subjective experience—is not considered to be a product of mind, it makes sense to wonder about the possibility of subjective experience in other systems that are not currently considered to be products of mind, at least as much sense as attempting to build a 'thinking machine' as an artefact, given the current state of ignorance about the mind-matter problem. From this perspective, if one accepts that the first living organisms— which seem to be a product of mind—appeared by virtue of the activity of the atmospheric machine, it becomes quite natural to pose the following question: is there subjective experience of mind related to the activity of such a machine? If so and if, in particular, there were a subjective experience of authorship, the role of the whole atmospheric system in the emergence of life would actually be analogous to that of a human brain in the emergence of a product of a human author's mind. In this case at least, the first well-formed living cell on Earth would have actually been a product of mind; that is, it would have come into being by virtue of nothing but the laws of nature in the third-person description, but to this description in the third-person perspective would correspond the designer's subjective experience of authorship in the first-person perspective, just as it happens for products of human authors. However, it should also be noted that, although the question about a possible subjective experience related to the activity of the atmospheric system may seem bizarre, in this case it would be easier to accept that a possible subjective experience of mental contents would have no causal power over matter; otherwise we would find physical events among lightning discharges that would not admit a physical explanation. When, instead, the question concerns the relationship between our brain and our mind, except for some desperate approaches to the problem, we have no problem admitting that we actually experience mental contents, but we typically find it much more difficult to accept that our mind has no true causal power over matter. By the way, let us also consider that today the best criticism against the 'Chinese room' argument still seems to be that the whole Chinese room does actually think in Chinese, since it—as a whole—seems to think in Chinese; likewise, in the classical scenario of the origin of life in which an apparent product of a thinking agent appears to be produced by the activity of the atmospheric system, it is quite natural to ask whether or not the whole atmospheric system thinks. These observations serve to emphasize that often what concerns the mind-matter problem is dotted with aspects that are difficult for a human to accept, but we certainly cannot exempt ourselves from investigating them because of it. Even if the story of the origin of life on Earth risks becoming even more personally unbelievable than those typically produced, we have to be aware that the argument from personal incredulity is a logical fallacy.

Therefore, once we have presented this scenario for the emergence of life on Earth, in order to rigorously assert that primordial organisms are not a product of design we should in fact prove that there has been no experience of authorship related to the discharge of the atmospheric system. As much as this possibility may seem fanciful, in the same way in which we have no way of proving that the cup-grabbing machine actually does not experience some mental content, we have no way of rigorously excluding it even in this case. And the reason why we can't prove or disprove it is that we don't know why the purposeless activity of neurons corresponds to the experience of our minds. Our position regarding the origins of life is therefore that, given the plausibility of the scenario proposed in physical terms for its origins, it is not trivial at all to state that primordial living organisms are not products of mind. As we have shown, the problem of the origin of life is inseparable from the mind-matter problem; thus the mind-matter problem cannot be considered a problem concerning, at most, all animal organisms and perhaps some human-designed machines, but rather it deeply affects the neo-Darwinian conception of the emergence of living organisms on Earth that is currently taken for granted. When Dawkins refers to a *blind and unconscious process* for the origin of living organisms, the point that is overlooked in his reasoning is that the blindness of the process depends on the descriptive point of view of the process: a system of neural discharges is always a blind process since forces of nature are blind by assumption, but a first-person experience of actions carried out with a purpose is related to such a purposeless process, however mysterious this still is for us. And because of this ignorance, it is not straightforward to exclude that the same could occur for a system such as Earth's atmospheric system, whose activity seems to have an important causal impact on the emergence of living organisms, which seem to be products of mind. Hence, with respect to the 'unconscious' attribute, to say that there is no experience of mental contents related to the activity of the atmospheric system, given the current ignorance about the mind-matter problem, is nothing but an unjustified belief, while knowledge typically requires at least a justified belief. Since we have no way of excluding that there is experience of mental contents related to the activity of the atmospheric system, we therefore choose to explore the possibility that the first living organisms are what they appear to be: a product of mind, and that, given the causal role attributed to the terrestrial atmospheric system, there is in fact subjective experience of authorship related to the activity of this system. In other words, we note that we are more justified in believing that there is subjective experience of mental contents related to the activity of the atmospheric machine—since it is believed to have a decisive causal role in the appearance of living systems, which strongly seem to be the product of a mind—than to believe that there is not. Basically, we are resuming the watchmaker argument of Paley, but we are referring it to a natural machine, rather than to a supernatural theistic God, which does not make much sense for a natural scientist, and we wonder if it makes sense to look at Earth's atmospheric system as the 'brain' of the Earth, i.e. a physical system whose activity corresponds to first-person experience. In this approach the author of life would be identified with the Earth, thought of as a subject of experience, but this perhaps arbitrary subjectification is not necessary: in this scenario it is sufficient to ask whether the activity of the atmospheric system could correspond to a first-person experience, particularly experience of the intention to realize life, without specifically identifying such a subjective entity. However, the designer may seem at first glance identifiable with our planet, thought of as a subject, possibly endowed with a 'brain', in the sense we just mentioned.

Such an approach could lead to placing this work within the subcommunity of biologists which constitutes the Intelligent Design movement, but this is not the case for the following reasons. Advocates of Intelligent Design contest the creative potential of the random mutation and natural selection mechanism, claiming instead the necessity of a designer to explain the origin of the species, neglecting a deeper discussion around the concept of authorship, while admitting the chance of a supernatural author, which is why the debate between the neo-Darwinists and the Intelligent Design movement is often connected with a theistic-atheistic debate. Our approach is completely different. As you can see, we are not criticizing neo-Darwinism as a scientific theoretical framework, but rather we are proposing a distinct interpretation of the framework, just as different philosophical interpretations of the same well-accepted theory of quantum mechanics are possible. Indeed, as natural scientists, we are interested in a naturalistic explanation of the origin of living organisms and we assume that the whole neo-Darwinian framework is valid for explaining the origin and evolution of living organisms. By the way, if some aspects of the framework need to be reviewed, we believe that it will always be possible to revisit them in order to provide a better explanation of the origin and evolution of organisms ultimately in purely physico-chemical terms. Leaving aside the theistic-atheistic debate, we point out that, anyway, a conclusion regarding the actual origin of living organisms on Earth is inseparable from the mind-matter problem, which is the problem we are ultimately interested in: we wish to show at least to what extent it pervades our knowledge and, above all, our ignorance, far more than is currently thought.

Once the terms of how we set up the problem are clarified, the point is that we do not know what makes the experience of a mind correspond to the activity of a physical system such as a human brain; but, as we mentioned in the previous chapter, following Searle's approach, for subjective experience to be caused by a physical system this must duplicate the causal power of the brain of causing the mind, and therefore we have conjectured that, in order to duplicate this causal power, the system must be somewhat similar to the brain both as a physical system in terms of its physics and as a computing machine in terms of its architecture. However vague this conjecture is, the best we can do at the moment is to undertake a brief comparative study between a brain and the atmospheric system to see if there are similarities between the two systems beyond the obvious differences. Certainly, in this analysis it should be considered what the atmospheric system could have been like on primordial Earth, but we refer for simplicity to the current one—although it is reasonable to suppose that the primordial system had to have a greater distance from the equilibrium, which gave rise to greater structuring and more intense activity.

In this regard it must be said that, as far as we know, in the scientific literature the question regarding a possible subjective experience related to the activity of the atmospheric system has not been explicitly addressed so far, but there is instead a long tradition of qualitative and quantitative comparative studies between the atmospheric system and the brain—which we will refer to—aimed essentially at the search for possible interactions between environmental electromagnetism and brain activity.

Certainly, the brain and the atmospheric system are both self-organizing complex non-isolated systems at thermodynamic disequilibrium. On the matter of complexity, we are not interested in quantifying it according to some arbitrary metric, while with respect to self-organization we have already mentioned several cloud patterns that can be framed in the context of dissipative structures. In other words, in the context of the debate we have mentioned about the relevance of dissipative structures in relation to the origin of life on Earth, we are exploring the possibility that they are actually significant, but in an indirect sense, namely that the subjective experience of mind could be related to a certain activity of such a complex self-organizing disequilibrium system. We argue, in this regard, that—since a computational theory of mind has been ruled out in the previous chapter—it is less hazardous to hypothesize subjective experience of mind in non-biological systems which share the property of self-organization with the brain than to hypothesize it in a Turing machine, which is of course not a self-organizing system, since we cannot exclude the idea that the property of self-organization might be crucial for subjective experience to happen. Considerations of this type clearly apply to all the similarities that we are about to discuss in terms of the physics and architecture of the two systems.

Certainly these are two systems whose main constituent is water, and both have a cellular structure at the lowest organization level (fig. 3.2a): in the scenario we have referred to for the origin of life, cloud water droplets are the most ancient ancestors of living cells and, therefore, also of neurons.

As concerns the global organization of the two systems, let us just note that the atmospheric system shows at the highest hierarchical organization level an organization in two hemispheres (fig. 3.2b)—characterized by typically opposite values of Coriolis force as a consequence of opposite latitudinal motion ultimately due to opposite latitudinal gradient temperature—as a sufficiently developed brain also shows. In other words, the two systems share a symmetry, which is the bilateral symmetry.

Furthermore, in these two non-isolated systems, free energy—which for both systems ultimately comes from the Sun—is involved in establishing electromagnetic potential differences across the membrane of every cell of the system. In the case of the brain, the electrochemical membrane potential is maintained by virtue of the action of the ATP-dependent Na-K pump, while for the atmospheric system part of the free energy absorbed by the Earth is stored in potential differences within a cloud, between clouds, between a cloud and the surrounding air, and between a cloud and the ground. Even though the mechanism responsible for charge separation is still debated and not yet fully understood, it is nevertheless reasonable to presume that they may mainly involve the electrostatics of convective motions, rather than electrochemistry.

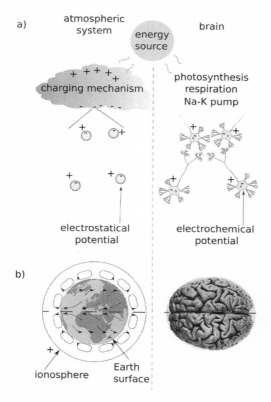

Figure 3.2 Comparison between the atmospheric system and the brain: a) Low-level cellular structure; b) Overall bilateral symmetry

Regardless of the specific charge-separation mechanism, because of the cellular structure of each cloud it is likely that a potential gradient is present across the surface of the cells of the system, which can generally consist of liquid water droplets or some form of solid water crystals (fig. 3.2a).

For both systems, the free energy absorbed and stored in such potential gradients is dissipated in the process of potential discharge, which consists in an electrostatic discharge for the atmospheric system and an electrochemical discharge for the brain. Cloud charging is the phenomenon responsible for lightning discharge events; most of such events are within clouds or between clouds, but cloud–air and cloud–ground lightning also occur. The physical mechanism through which a lightning discharge begins is the so-called 'stepped leader', which is constituted by a channel of free electrons proceeding from the negatively charged region of the cloud in discrete luminous segments of tens of metres in length, typically branching in a tree-like configuration. When the stepped leader approaches the positively charged region of the cloud, its relatively large negative charge sets the condition for the 'return stroke', a large flow of positively charged ions that proceed from the positive region to the negative region of the cloud. At the level of the cells of the system, what happens is that, when the negatively charged cells synchronously discharge, they trigger—through the stepped leader current—the synchronous discharge of the positively charged cells, where the overall process expressed in these terms resembles the synchronously induced discharge of a neuron cluster by virtue of the discharge of another neuronal cluster (fig. 3.3a). In the context of the comparative literature about the two systems, Persinger (Persinger 2012) worked specifically in comparing the saltatory conduction in neural axons with the stepped leader. He also recognizes similar wave-shape characteristics between the action potential and lightning discharge, and the coincidence of different physical quantities calculated for a brain and for the atmospheric system, such as the density of power of the discharges ($10^{-7} W / m^3$) and the total current density ($10^5 A / m^2$). Regardless of the quantitative coincidences found by Persinger, a lightning discharge in which the intracloud potential is discharged can be seen as an event in which the potential gradient across the surface of a population of droplets suddenly vanishes or generally changes and, because of the return stroke, a potential gradient is induced across the surface of the cells that make up the opposite electrode within the cloud. From this point of view, the lightning current constitutes an analogue of the axonal currents of two clusters of connected neurons, although in this case the process is electrostatic rather than electrochemical. As regards the specific difference between the two processes, in neurons the discharge occurs when the potential goes below a certain threshold, while in the case of cloud droplets, conceived as the neural cells of such a brain-like system, the synchronous discharge of cells occurs when the potential exceeds a certain threshold, since dielectric breakdown must occur. Moreover, once a lightning discharge has occurred, if additional negative charge is made available to the upper portion of the previous stroke channel in less than about 100 milliseconds from the cessation of the current of the previous stroke, when this additional charge is available, a continuously propagating leader, the 'dart leader', moves down along the defunct return-stroke channel, again depositing negative charge from the negative charge region along the channel length. The dart leader thus sets the stage for the second return stroke, and the process can be iterated as far as the conditions for breakdown are maintained. The result is a train of impulses (fig. 3.3b) produced by the second cell cluster stimulated by a train of impulses coming from the first cluster. When the time interval between them is sufficiently large, they give rise to the typical strobe-light effect that is observed in thunderstorms. Similarly, trains of impulses in synchronously firing neurons generally occur in the brain.

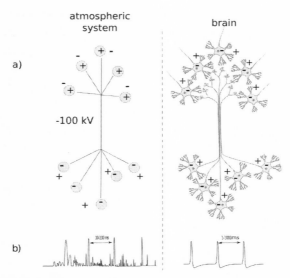

Figure 3.3 Comparison between the atmospheric system and the brain: a) Paths of currents occurring because of cellular potential discharges; b) Pulse trains

At this point, it is natural to ask whether atmospheric discharges can be related events, just as distinct groups of synchronous discharges of neuron clusters are related in a brain. In other words, we are asking whether the atmospheric machine can be seen as a computing machine that presents an architecture conceivable as a neural network of some kind. In this regard, while an enormous number of studies have been devoted to the initiation of a single lightning flash, the characteristics and the properties of a sequence of consecutive flashes have received little attention, most likely owing to the inherent assumption that a lightning flash is a stochastic process, unrelated and independent of the occurrence of other flashes in the same thunderstorm. Dennis (Dennis 1970) visually recorded 20 thunderstorms in New Mexico and conducted a statistical analysis on the timing of consecutive flashes: he concluded that the occurrence of an individual flash must be considered a random phenomenon. But the reliability of observations and the limited number of analysed cases have suggested that there may be a different way of interpreting the interarrival times of flashes. In a different study by Mazur (Mazur 1982) of lightning in different electrically active thunderstorm regions,[1] many cases were reported of 'associated discharges', namely flashes occurring within a 200ms interval from one another. Mazur studied the null hypothesis that all observed flashes were independent events and showed that the null hypothesis is rejected at a significance level between 0.1% and 5%. The physical interpretation of this interdependence was based on the electrical interaction between neighbouring active regions, where the collapse of an electric dipole in one region (due to the discharge) induces a sudden change in the electric field vector in a neighbouring region, although the nature of this electromagnetic coupling does not seem to be completely clear yet. A similar phenomenon of closely spaced flashes was reported by Vonnegut et al. (Vonnegut et al. 1985), who noted the clustering or convergence of lightning flashes as they appear in nocturnal space-shuttle video images. They just offered a conceptual explanation similar to that of Mazur, stating that the release of electrical energy in one portion of the cloud eventually triggers the breakdown process in another part. In another study using space-shuttle lightning images recorded in a previous study (Yair et al. 2004), Yair et al. analysed footage from six observed storm systems (Yair et al. 2006), and showed that in storms exhibiting a high flash rate, lightning activity in separate electrically active regions showed transient synchronization with bursts of nearly simultaneous flashes in different regions. They proposed that this behaviour is equivalent to the collective dynamics of a network of weakly coupled limit-cycle oscillators. According to this hypothesis, thunderstorm regions embedded within a mesoscale convective system constitute a network, and the flash occurrence rate can be described in terms of phase locking of a globally coupled array. Comparison of basic parameters of the observed lightning networks with predictions of random-graph models revealed that they are best described by generalized random graphs with a prescribed—related to its activity—distribution (Newman, Strogatz, and Watts 2001), which typify networks supporting fast response, synchronization, and coherent oscillations. Yair et al. (Yair, Reuven, and Ravid 2009) also attempted to explain patterns of clustering and synchronicity in separate thunderstorm convective regions, at a distance of tens to hundreds of kilometres from each other, observed from various ground-based lightning location systems, confirming previous observations of lightning synchronicity based on space-shuttle images, and hinting at a possible mutual electromagnetic coupling of remote thunderstorms. In practice, from the observations it appears that lightning strikes even hundreds of kilometres away from each other are not independent events. Although the electromagnetic coupling mechanism is not yet fully understood in detail, the authors have modelled the phenomenon adopting a model borrowed from neuroscience: the adaptive network of leaky-integrate-and-fire (LIF) oscillators, a classical electrical model of a neuron or a synchronously firing neuron cluster. With respect to their work, we observe that if the breakdown of a cloud electric field and the subsequent lightning discharge is seen as a cluster of synchronously firing neurons, they attempted to explain the synchronous lightning discharges of remote thunderstorms in terms of synchronously firing neuron clusters.

In fact, the above-mentioned recent studies are part of a rather long tradition of comparative studies between the atmospheric system and the brain, whose beginning dates back to the 1950s. Another noteworthy element regarding atmospheric discharges is that they occur within a region that can be modelled as a spherical capacitor whose two armatures are made up of the Earth's surface and the ionosphere, where the capacitor is kept charged by lightning discharge to the ground, which participates in the global atmospheric electrical circuit (GEC), i.e. the course of continuous movement of atmospheric electricity between the ionosphere and the Earth (fig. 3.2b). The phenomenon of typical lightning that occurs in the troposphere (0–20 km altitude range) is in fact only a part of a wider electric breakdown phenomenon that extends to the ionosphere (beyond 70 km) where upper atmospheric lightning occurs, known as transient luminous events (TLE). These contribute to the charge of the upper armature of the capacitor, while a weak direct current that passes through the atmosphere — ionized by virtue of the continuous impact of cosmic rays — tends to continuously discharge the capacitor. The entire circuit is therefore in the first instance schematizable as a solar-powered electrostatic generator that charges a spherical capacitor, placed in parallel with a leaky resistor. In this regard, in 1952 Winfried Otto Schumann, at that time Director of the Electrophysical Institute at the Technical University of Munich, published his first paper (Schumann 1952) about quasi standing electromagnetic waves in the waveguide, which is formed by the Earth's surface and the ionosphere generated by lightning discharge that occurs in such a cavity, and reported the first observation of them in collaboration with Herbert König (Schumann and König 1954). König, who was a leading expert in the field on the effects of environmental electromagnetism on biological systems, noted remarkable congruencies between the waveforms of electroencephalographic activity recorded from the scalps of human subjects and the patterns of naturally occurring electromagnetic activity generated by global lightning. In particular, the Schumann resonances, which are traditionally defined by spectral peaks at approximately 8, 14, 20, 26 and 33 Hz (Balser and Wagner 1960), show striking consistency with electroencephalographic activity, both in terms of frequency and intensity. In this regard, in 2006 Pobachenko et al. (Pobachenko et al. 2006) reported evidence of real-time coherence between variations in the Schumann and brain activity spectra within the 6–16 Hz band for a small sample. The experiment was successfully repeated by Saroka (Saroka and Persinger n.d.) for a larger sample.

From this brief comparative discussion between the two systems we are interested in, it appears that, although the two systems are certainly different, they present various similarities in the aspects we have illustrated. Basically, it would be worth evaluating whether there is a case for asserting that a brain is a scaled-down version of the atmospheric system, based on electrochemistry rather than on electrostatics. Anyway, despite the elements that have emerged so far, we have no way of establishing whether there is subjective experience related to the activity of the atmospheric machine. Once the computational approach to the mind has been discarded and we have grasped the similarities between the atmospheric system and the brain, we have no way of determining whether the atmospheric machine can duplicate the same causal power of the brain to cause the subjective experience of mind, or, conversely, whether the brain does duplicate the specific causal power possibly possessed beforehand by Earth's atmospheric machine and reasonably by the atmospheric machines of other planets where some stuff that has the appearance of design could potentially be found. We can only conclude that, given the similarities in terms of the physics and architecture of the two systems, it may be interesting to actually continue to consider the possibility that there is or has been some subjective experience related to the activity of this system. Certainly, an in-depth comparative study would require much more space, and we believe it is a useful analysis to perform, but even a more in-depth study would end up including an in-depth report of similarities and differences in the physics and architecture of the two systems, from which it would still not be possible to conclude whether or not there actually is subjective experience related to the activity of this system. Furthermore, even if we could reach an affirmative conclusion on the matter, it would not directly imply an experience of authorship in the realization of life on Earth and, therefore, that life on Earth is the product of Earth's mind. So, if we want to further analyse the question we have to explore other paths.

Before proceeding further, it is useful to underline that here we are dealing only with the problem of the origin of life, while neo-Darwinism addresses the entire evolutionary history of the tree of life. In fact, we could equally question whether the atmospheric system had a relevant causal power in the development of species and make similar considerations concerning the mind-matter problem in the evolution of species. In this regard, König (König and Ankermüller 1960; König, Lang, and Krueger 1981) was the first to argue that such a stringent coincidence of parameters between the brain and the atmospheric system is an indication that environmental electromagnetism must have played a decisive role in the development of living organisms. Independently of König's observation, the question of the interaction between the entire dynamics of planet Earth and the evolution of species is certainly addressed in the context of the fascinating and controversial Gaia hypothesis by Lovelock and Margulis. Moreover, we are well aware now of the mutagenic power of many pathogens which can be found at high altitudes, and, as we mentioned, lightning activity is a source of ionizing mutagenic radiation; therefore it can't be excluded that the atmospheric system may have a more or less indirect action on the evolution of species. Anyway, even though there are tons of scientific papers that discuss the role of the atmospheric system in the origin of life, regarding the role of the atmospheric system in the evolution of species we do not have much more than vague considerations and controversial theories available; thus we choose to restrict this analysis about the impact of the mind-matter problem on the neo-Darwinian conception of the biological world exclusively to the impact of the mind–matter problem on the current conception of the origin of life on Earth, without discussing whether or not it is straightforward to exclude any experience of intentionality underlying the emergence of new, more complex species.

Once the problem is confined to the emergence of life on Earth, our question is the following: if a description of the origin of life in terms of authorship is not actually necessary in a materialistic perspective — as it is never necessary in any case — how do we explore the possibility that such experience of authorship might in fact exist?

Certainly, the more direct path we can think of — since the atmospheric system is still here — is to attempt to communicate with this system provided that there is still subjective experience related to its current activity, posing to 'them' the question we are interested in. Such a route might sound like science fiction, and brings to mind what happens in Fred Hoyle's novel *The Black Cloud*, in which humans communicate with an intergalactic intelligent cloud. We believe that such a route would not be any more desperate than communication attempts currently performed with the SETI experiment, but we still believe that attempting such a communication would be premature in the absence of other clues about the issue.

Obviously, the typical way of attempting to establish whether a machine should at least be considered as a subject of experience is to perform some version of the Turing test, but the original formulation of the Turing test, as in each of the countless subsequent variations, involves the existence of human–machine communication, which is precisely what we don't know how to achieve with the atmospheric machine, since we are obviously not the designers of such a machine. Therefore, what we can do is to borrow the principle underlying the Turing test and apply it, as far as possible, to the scenario we are contemplating. This underlying principle is the imitation principle: if the behaviour of a machine imitates a bit of human behaviour to the point of being indistinguishable from such a behaviour, then the machine must be considered as a subject of experience, in the same way that we consider our fellow humans as subjects of experience. The behaviour actually resides in the responses to the stimuli that the system receives from the environment, and what makes a behaviour appear human-like is the meaningfulness of the response with respect to stimuli. Basically, if the response to stimuli we receive from the machine in the Turing test is meaningful to us as if we had given it, given that we did not actually give the response, it is reasonable to think that it is significant for the one who did actually give it, and therefore they should be considered as a subject of experience — experience of authorship in this case — in the same way as our fellow humans. So, in this analysis we are tempted to explore whether the emergence of life on Earth can be conceived as a meaningful production by an author, that is, as a meaningful response of the planet to certain environmental stimuli, a response that a human would equally — at least in principle — produce. In this case the machine in question is the atmospheric machine, and if we want to apply the above testing principle to it we must explore which stimuli give rise to the activity of the machine.

So far, in the analysis of the atmospheric system as a 'brain', we have not discussed how reception might work; that is, we have not considered the possible causal power of physical stimuli coming from the system's environment over the system's activity, i.e. the production of discharges and discharge patterns, which is basically what happens in a brain, for example, when visible photons hit retinal receptors. We have seen that discharges in the atmospheric system are interrelated events as in the brain; that is, in this system some non-trivial form of computation takes place, in the same sense in which it occurs in the brain, and we have also seen how the system can act on proto-living systems — such as coated water droplets — to realize the first living systems. What is missing therefore is a study of the way in which the system could be stimulated within its environment by physical events. In other words: is it possible that some discharges are induced by something physical coming from the system's environment?

Despite being one of the most familiar and widely recognized natural phenomena, lightning discharges still remain relatively poorly understood (Dwyer and Uman 2014). The main problem of lightning physics is that the electrical fields that occur between the clouds (10–100kV/m) are typically an order of magnitude lower than those required for dielectric breakdown of the atmosphere (2MV/m); thus, the physical mechanism that initiates many lightning strikes is not yet completely clear. Furthermore, as we mentioned, during thunderstorms, X-ray flashes ($50keV$) and γ-ray flashes ($0.05-10MeV$) are detected. Gurevich (Milikh and Roussel-Dupré 2010) argues that the existence of high-energy emissions indicates that relativistic electrons must play a significant role in thundercloud discharge; therefore the initiation mechanism proposed is a relativistic runaway electron avalanche (RREA). The physical process proposed for the initiation of lightning is the generation of new fast electrons from the runaway-particle ionization of neutral molecules. Although the majority of newborn free electrons have low energies, some will have energy above a certain critical threshold. Those will also be accelerated by the cloud electric field, become runaway electrons, and may in turn generate more free electrons of sufficiently high energy. As a result, an exponentially growing avalanche of accelerated electrons can occur. The electric field required for the relativistic breakdown phenomenon is actually an order of magnitude lower ($200kV/m$) than the one required for the classical breakdown. In this scenario, however, reaching the critical field threshold is not the only necessary condition for the breakdown to occur.

The presence of fast 'seed' electrons, having energies above the critical runaway energy of $100keV$, is also necessary. Therefore, a runaway breakdown can be stimulated by fast 'seed' electrons provided by an extensive atmospheric cosmic ray shower. The altitude of the clouds is in fact below the altitude where the cosmic rays begin to interact significantly with the atmosphere: many cosmic rays interact when the pressure is 100–200hPa, therefore at an altitude of 11–16 km in the upper troposphere or in the lower stratosphere. In this approach, therefore, cosmic rays play a decisive role in initiating atmospheric discharges, where clouds (in peculiar conditions) basically constitute natural versions of spark chambers. This is interesting because in this case Earth's atmosphere would constitute not only a natural target for cosmic rays, but also to a certain extent a detector; that is, a particular lightning discharge would constitute for an external observer a sign of a cosmic ray impact, a natural detector in terms of physics, or a natural receptor in neurobiological terms. If we admit that these discharges directly induced by these events may induce others in turn, the whole process of discharging is analogous to what happens in our brain when visible photons impact our retinas, and we know that in relation to such brain activity we experience a visual perception of an object or event. Therefore, if we admit the possibility that there is subjective experience related to the atmospheric discharges we are dealing with, we can infer that the content of this experience will be some form of perception of the cosmic ray event. This type of perception is obviously very difficult for us to imagine, at least as difficult as it is to imagine what it is like to be a bat — whose perception mechanism based on sonar is very different from ours — as pointed out by Nagel in his famous work.

We should at this point attempt to verify whether or not the behaviour of Earth's atmospheric system, in which the production of life is thought of as a response to environmental stimuli—cosmic ray events—satisfies the imitation principle, i.e. whether or not the response to stimuli is meaningful for a human as if it were given by a human. As we can see, dealing with the two problems—the mind-matter problem and the origin of life—in a joint way introduces the possibility that the phenomenon of life on Earth still needs further exploring and that it might admit a semantic plane of understanding, which would at least be a clue for the possibility that it is actually a product of mind. In other words, it is a question of establishing whether the atmospheric system is a machine which, when subjected to natural stimuli (cosmic ray events), produces a significant response to these (life on Earth) just as a human brain would do. Hence, we have to verify whether it is possible to find a semantic plane of understanding of what life is as a meaningful response with respect to stimuli that are constituted by cosmic ray physics.

Therefore, similarly to the brief overview presented in Chapter 1 on the simplest living organisms, in the next chapter we briefly summarize some key elements of the physics of cosmic rays, with the aim of finding some clue that can contribute to an understanding of what life is on a possible semantic plane of comprehension, admitting the possibility that life could actually be what it appears to be: a product of mind, and, in this specific scenario, the product of a mind whose mental contents would consist of elements of cosmic ray physics.

Chapter 4

The Seeds of Life in Cosmic Ray Physics

The composition of primary cosmic rays which come from outer space and hit Earth's atmosphere—as they hit of course any other celestial body—is the following:

- 89% of them are protons;
- 9%—α particles;
- 1%—heavier nuclei;
- 1%—electrons;
- $\ll 1\%$—γ-rays.

Regarding the origin of cosmic rays of a given energy E:

- particles with energy $E < 10^{10}\,eV$ come mainly from the Sun;
- $10^{10}\,eV < E < 10^{16}\,eV$ —they are attributed to sources in our Milky Way Galaxy; supernova explosions, which implement the famous Fermi acceleration mechanism, are regarded as the best candidates;
- $10^{16}\,eV < E < 10^{18}\,eV$, between the knee and the ankle of the spectrum—their origin is unclear since they are again thought to be produced within the Galaxy, but the energies are too high for them to be accelerated by the shocks of supernova remnants;
- $E > 10^{18}\,eV$ —they are thought to be of extragalactic origin: they may come from active galactic nuclei, from two merging stars, or from a star merging with a black hole. The interaction of such extragalactic cosmic particles with the cosmic microwave background sets a limit to their energy of about $5 \cdot 10^{19}\,eV$, the GZK limit, which corresponds to

a centre of mass energy $\sqrt{s} = \sqrt{2Em_p} = 10TeV$, which is comparable to the centre of mass collision energy of the Large Hadron Collider (LHC).

Such cosmic rays, by interacting with a nucleus of the atmosphere (mainly nitrogen nuclei), lead to the production of other particles, which in turn can interact with atmospheric nuclei, leading to the production of other particles. As long as such products are energetic enough, the process can be reiterated, and the result is a 'shower' of particles, which is typically referred to as an *air shower*.

For example, when a photon passes through the Coulomb field near an atomic nucleus of the atmosphere, an electron and a positron can be created. Under the same conditions, the electron and the positron undergo *Bremsstrahlung*, radiating photons; the cascade of repeated collisions leads to a shower of electrons, positrons and photons, an e.m. shower.

Showers are far more likely to be caused by a proton or a heavier nucleus. When such a cosmic ray particle enters the atmosphere, a hadronic interaction will occur with an atmospheric nucleus. The collision of these particles with nuclei of the atmosphere leads to the typical process of production of hadron jets by parton fragmentation, and most of them are pions, the lightest hadrons. The neutral pions, with a mean lifetime of $\tau = 8.4 \cdot 10^{-17} s$, decay almost instantly into two γ particles

$$\pi^0 \rightarrow 2\gamma \left(BR = 98\% \right)$$

giving rise to electromagnetic subshowers, which in turn can initiate the RREA mechanism mentioned in the previous chapter. The charged pions, with a mean proper lifetime $\tau = 26ns$, can collide with other nuclei, generating new pions, which then populate such a hadronic shower. When the energy of a pion is not large enough for further pion production in the next collision, it will decay into a muon and a neutrino. Muons will either survive to the surface of the Earth or decay into an electron — which in turn can induce e.m. subshowers — and two neutrinos.

At high collision energies the process of further production of hadrons takes place through the typical parton shower and subsequent hadronization, while at low interaction energy the process can basically be described as a breaking up of the incoming hadron into two or more hadrons. Such a process also occurs when, for instance, a high-energy cosmic photon (a γ-ray) hits an atmospheric proton and breaks it apart into a nucleon and a pion:

$$p + \gamma \rightarrow p + \pi^0, n + \pi^+$$

Equivalently, at very high energies, a proton can collide with a low-energy photon such as that of the cosmic microwave background (CMB), giving rise to the process that is responsible for the GZK limit. In the proton rest frame, the process can be described in this way: the proton breaks up into two or more hadrons by absorbing available energy from its environment; in this case, the absorbed energy is that of a CMB photon which in the rest frame of the proton appears as a high-energy photon. The process can involve the formation of a resonance, such as, in this case, the Δ^+ or even the Δ^{++} resonance, which finally breaks up into the final-state hadrons. Indeed Taylor, in his famous book *Scattering Theory* (Taylor 1972), provides the following heuristic definition of resonance as: "a long lived state of a system which has sufficient energy to break up into two or more subsystems". Like any other process of energy absorption (by a nucleus, an atom, a molecule), the formation of the resonance involves energy-mass transformation; but, in this case, the fate of resonance is very different from the typical re-emission of the absorbed energy: the break-up of the resonance involves the actual creation of a particle-antiparticle pair, and, given that we are dealing with quarks and antiquarks, these are confined in the 'subsystems' into which the original system is fissioned. Furthermore, the subsystems belong to the same class that the original system belongs to; that is to say they are still hadrons, so that the process of production of hadrons by breaking up a hadron can, in line of principle, be indefinitely iterated, regardless of the fact that the system passes through a resonance. And it is in fact iterated in the repeated collisions of the cosmic protons with the photons of the CMB, as well as in low-energy collision in hadronic showers,[1] for which the following description in the hadron rest frame is available: hadrons absorb energy from their environment and, by virtue of such energy absorption, they undergo a break-up process which leads to the production of further hadrons.

We therefore observe that each system belonging to the class of hadrons enjoys this particular property: when such a system absorbs energy from its environment, other systems of the same class (other hadrons) are produced by fission of the original system. Since the products belong to the same class as the original system — because they are hadrons too — the process of fission through energy absorption can, in principle, be indefinitely iterated. We can't avoid noting that, because of this property, the peculiar behaviour of hadrons absorbing energy from their environment allows the following description in terms borrowed from biology: if the initial hadron is thought of as the parent hadron and the products of the fission are thought of as the progeny, the characteristic property of the class enjoyed by both progeny and parent can be thought of as inherited by the progeny from the parent; in other words, a system belonging to the zoo of hadrons gives rise to other systems of the zoo by absorbing energy from the environment, and, as long as there is energy available, the process is iterated, as happens in hadronic showers. It could be argued that inheritance in biology makes sense in the context of a description in terms of hereditary information, which for hadrons is meaningless. The point is that, as we said, for living organisms the description in terms of information is a useful subject-relative description of a biochemical process, that is, of a process ultimately governed by nothing other than the laws of electromagnetism. Therefore we find it interesting that this notion of inheritance may in fact be applied not only in complex systems governed by the laws of e.m. — living organisms — but equally, albeit in an absolutely minimal form, in the simplest systems of elementary particles ruled by the laws of strong interactions: hadrons.

Consequently, in this sense, hadron showers are somewhat reminiscent of a thriving population of elementary organisms that survive as long as there is enough available energy in the environment, reproducing by fission. The constituents of the shower are mainly mesons (pions), which are indeed metastable objects that finally decay into leptons and photons (objects that do not belong to the class of hadrons). Therefore, we could venture towards the following description of the process under a biologically biased perspective: pions need to absorb energy to reproduce by fission — i.e. they break apart, producing daughter pions — otherwise their existence comes to an end by decay, as happens at the end of a hadronic shower.[2]

Regardless of how we want to describe the process, once the available energy for the further production of hadrons runs out, the cascade process stops. By the way, as mentioned in the previous paragraph, atmospheric electric fields can actually further accelerate secondary cosmic ray particles, but even though these fields can trigger a relativistic runaway electron avalanche, their effect is almost irrelevant on the charged pions of a hadronic shower, since they are too heavy compared to electrons. It is, however, useful to briefly consider an artificial scenario in which the charged pions produced in an initial collision are actually accelerated while crossing a sufficiently dense medium. The pions are of opposite charges; therefore we assume that the π^+ are accelerated while the π^- are slowed down. We can dismiss them since they do not survive the slowdown and eventually decay into muons. So we assume that in each collision a $\pi^+ \pi^-$ pair is produced, of which the π^+ is further accelerated and the π^+ decays due to the slowing down, so the population of π^+ increases by 1 at each collision. For the loss of energy due to the interaction, the average energy of the π^+ will reach a constant value depending on the mean free path and the power of the accelerator. In this scenario, the trend of the population of π^+ is due to the following concurrence between decay and growth:[3]

$$dN = \left(\beta c n \sigma - \frac{1}{\gamma \tau} \right) N dt$$

$$N(t) = N_0 e^{\left(\beta c n \sigma - \frac{1}{\gamma \tau} \right) t}$$

where

$$\beta \gamma c \tau = \frac{1}{n \sigma}$$

—that is, the equality of the mean path length $\beta \gamma c \tau$ and the mean free path $1/n\sigma$ constitutes the critical condition for the trend of the population. Therefore, for these objects too, as for living organisms, the exponential growth and decrease of the population concur under the appropriate conditions; this is actually the case in every hadronic shower, although such showers inevitably tend to run out since—contrary to this artificial scenario—there is no constant energy input.

In light of what has been seen, it is useful to compare the behaviour of mesons with that of other particle-antiparticle systems. The two fundamental interactions of particle physics that form bound states are the strong force and the e.m. force, but although the e.m. interaction can constitute bound states, lepton-antilepton systems, such as positronium, do not enjoy the same property in regard to energy absorption as a quark-antiquark system; that is, supplying energy to a positronium leads to nothing but its excitation or its ionization. Therefore the behaviour of hadrons with respect to energy absorption is linked to the fact that the strong interaction can not only form bound states (nuclei), just like the e.m. interaction (atoms, molecules, intermolecular bonds), but—unless particularly exotic conditions are reached—it necessarily forms bound states between elementary particles (hadrons as quark systems). It is therefore useful to mention two fundamental characteristics that distinguish the strong interaction from the e.m. interaction.

QED   QCD

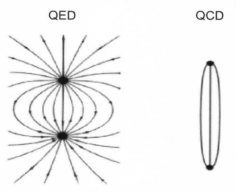

Figure 4.1 Field lines of the strong interaction field compared to field lines of the e.m. dipole field. The reason for the different field configurations is the self-interaction of the strong interaction field, related to confinement

1. *Quark confinement*. This was originally proposed to account for the fact that no single free quark has been observed in experiments: quarks are typically clumped together in triplets (baryons) or pairs (mesons). Confinement is currently believed to arise from a fundamental characteristic of the strong interaction, which is the self-

interaction of the strong interaction field (with three and four gluon vertices). Strong field self-interaction would be what makes the strong interaction linearly increase with distance (F α- r) instead of following the inverse square which describes the e.m. dipole interaction ($F \propto -1/r^2$); this feature is often illustrated by comparing the field lines of an electric dipole with those of the gluon field inside a meson (fig. 4.1).

2. *Quark asymptotic freedom.* Quarks move quasi-freely at very high energies compared to the rest mass of protons, and the interaction between quarks inside a hadron, characterized by an effective ('running') coupling constant, monotonically approaches zero; therefore, perturbation theoretical tools, well known from QED, can be successfully applied in this regime. This theoretically well-established consequence of QCD has also been verified experimentally.

One of the most notorious models for accounting for these features is the MIT 'bag model'. Developed in 1974 at the Massachusetts Institute of Technology shortly after the formulation of QCD, it soon became a major tool for hadron theorists. According to the simplest version of the model, quarks are forced by a fixed external pressure to move only inside a given spatial region, as if they were in a bag. Within this region, quarks occupy single-particle orbitals similar to nucleons in the nuclear shell model, and just like in nuclear physics the shape of the bag is spherical if all quarks are in the ground state; the bag results as stable if and only if the total colour-charge vanishes (colour confinement). The bag is considered to be physical and effectively generated because the quarks effectively push out all non-perturbative gluons from the inside of their spatial region, and therefore the radius of the bag reflects an equilibrium configuration: the pressure from the outside due to the non-perturbative gluonic interactions balances the pressure from the inside due to the motion of the quarks within the bag; since there are no real gluons present in the inside of the bag, quarks move almost freely in this region. Such a model relates in a plausible way to mechanisms that are known to be mechanisms of QCD, and this theory is, after all, supposed to be the fundamental theory in this domain. Therefore, the bag model is considered to be a plausible description of a hadron: something like this is probably what effectively happens inside a hadron. Thus, given the potential energy of the bag of a hadron of radius R

$$E_p \propto R^3 B^4$$

where the term B^4 is due to the energy density of the bag in order to fit the mass and radius of the proton, the bag constant results

$$B \approx 200 MeV$$

Obviously, although the bag delimits a hadron as a physical system by confining quarks, it is permeable to an energy flow. For example, it is possible to transfer energy into a hadron to obtain a higher-energy resonance, subsequently fissioned into two or more subsystems through the splitting of the bag, which in this sense seems to constitute the semipermeable membrane of a hadron, letting energy enter the hadron, but preventing quarks from escaping. Regardless of the formation of a resonance, the splitting of the bag allows the hadrons, by absorbing energy from the environment, to reproduce by fission; i.e. the splitting of the bag allows the daughter mesons to be produced by fission of the parent.

Something like this occurs in the low-energy interactions of the hadronic showers; specifically, at low energy the hadrons produced are basically mesons (mostly pions). It is interesting to note that mesons constitute a subclass of hadrons which reproduce by absorbing energy, as long as the transferred energy remains below a certain threshold, and, unlike baryons, they are all metastable. Furthermore, as previously seen in the scenario of artificially accelerated pions in a sufficiently dense medium, energy supply seems to be vital to sustain a population of pions, since pions are metastable objects that ultimately decay into objects not belonging to the class of mesons (leptons and photons). If, instead, a given amount of energy necessary for the meson fission is transferred into the system to push quarks apart, and a little fraction of the energy provided (typically less than 5% in mesons) is transformed into the mass of a newly created quark-antiquark pair because of its confinement property—where the newly created quark binds to the parent meson antiquark, and the newly created antiquark binds to the parent meson quark—then the splitting of the bag ultimates the fission (fig. 4.2).

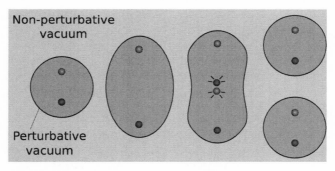

Figure 4.2 Binary fission process of a meson which absorbs energy from its environment

Given this interesting behaviour of hadrons and in particular of mesons, the question we ask in the context of this specific investigation is this: is it possible that a specific hadron undergoing fission by absorbing energy from the environment could give rise to two or more mesons identical in the final state to the initial hadron, similarly to how a single-celled organism splits into two children organisms genetically identical to the parent? The answer to our question is in principle affirmative. The process we are interested in is in principle possible for all mesons identical to their own antiparticle, such as the neutral pion π^0, which is present in secondary cosmic rays, even if in practice its average life is so short that it cannot reproduce itself in a typical shower but decays immediately into two photons, giving place to an e.m. subshower. Thus, only charged pions participate in the process of pion reproduction in the atmosphere. Postponing the discussion of this last point, let's begin by mentioning some well-known processes in which the neutral pion effectively reproduces itself by absorbing energy.

The neutral pion is a pseudoscalar meson; thus it is characterized by the quantum numbers $J^{PC} = 0^{\mp}$, and belongs to a subclass of mesons which are identical to their charge-conjugated system, such as the pseudoscalar mesons η, η' and the more massive mesons η_c, η_b. For the conservation of quantum numbers to occur, it can undergo reproduction, for example, in an interaction such as:

$$\pi^0 \pi^0 \to \pi^0 \pi^0 \pi^0 \pi^0$$

in which it reproduces itself by interacting with itself. Another interesting aspect regarding the neutral pion is that it is the lightest hadron. So, if the centre of mass energy $\sqrt{s}$ of a $\pi^0 \pi^0$ collision is so fine-tuned:

$$2m_{\pi^0}c^2 < \sqrt{s} < 2m_{\pi^\pm}c^2$$

$$270 MeV < \sqrt{s} < 280 MeV$$

then, unless elastic scattering occurs, they must necessarily reproduce themselves according to the previous reaction, because the production of a couple of charged pions would violate the energy conservation law. Alternatively, neutral pion reproduction can occur by electron scattering on a neutral pion

$$e - \pi^0 \rightarrow e - \pi^0 \pi^0 \pi^0$$

$e - \pi^0 \rightarrow e - \pi^0\pi^0\pi^0$

in which two more pions appear in the final state; the fine-tuning condition on the kinetic energy of the electron E in the pion rest frame is the following:[4]

$$4m_{\pi^0}c^2 < E < 2m_{\pi^\pm}\left(1 + \frac{m_{\pi^\pm}}{m_{\pi^0}}\right)c^2$$

$$540 MeV < E < 570 MeV$$

Given that the neutral pion can reproduce itself by absorbing energy from its environment, it is interesting to have a look at the reproduction process from a biologically biased perspective. By absorbing energy, such a meson is able to transform part of the incoming energy into the mass of the newly created quark-antiquark pair. As we have seen, all metabolic processes of biological organisms are characterized by energy-mass conversion, and in particular energy-to-mass conversion mostly occurs for anabolic processes and, conversely, mass-to-energy conversion mostly occurs for catabolic processes. If we want to look at the neutral pion as a living organism, we have to consider that this form of life is based on the strong interaction rather than on the e.m. interaction; therefore the quantities of energy converted into mass and vice versa will be several orders of magnitude greater than those that occur in the chemistry of living organisms. But, regardless of the quantitative difference, the process of converting a fraction ($2m_q/m_\pi \approx 5\%$) of the absorbed energy into the mass of the quarks necessary for the reproduction of the object can be considered, for all intents and purposes, as an elementary anabolic pathway in which the pion autonomously synthesizes what it needs for its reproduction.[5] If the valence quark structure of the pion is preliminarily thought of roughly as a quark-antiquark pair, each of the two newly produced quarks will bond to each of the respective antiparticles of the parent pion; the quark structure of each of the child pions is therefore constituted by a quark of the parent hadron bound to its (autonomously synthesized) respective antiparticle. In biological terms, the replication process of the pion quark structure is therefore a basilar semi-conservative replication, such as that of the DNA of living organisms. Once the semi-conservative replication of the quark structure has been completed, the bag—the pion's semipermeable membrane—is split in two, ultimating the reproduction process of the pion. Clearly, in this process the daughter particles enjoy the same property as the parent meson since they are identical to it— they are able to reproduce themselves by absorbing energy from the environment—so that such a property can be thought of as inherited from the parent by the progeny. The process can in principle be indefinitely reiterated, giving rise to a scenario in which, in line of principle, exponential growth and degrowth compete according to the resources available in the environment in which the pions live, just as happens for typical living organisms.

We must now make a necessary clarification regarding semi-conservative replication, because the description of the quark structure of a neutral pion as an antiquark-quark pair is clearly very crude. The physical state of the system is in fact a superposition of states, and, leaving out the space wave function, it is given by the product of a singlet of SU times the spin-flavour wave function, which is not a pure state due to the approximate symmetry SU. In this regard, although as we said the description of the process in terms of hereditary information here does not make sense, it is still useful to observe that, given the spin-flavour-colour wave function of the pion

$$\pi^0 = \frac{u\uparrow \underline{u}\downarrow - u\downarrow \underline{u}\uparrow - d\uparrow \underline{d}\downarrow + d\downarrow \underline{d}\uparrow}{\sqrt{4}} \otimes \otimes \frac{r\underline{r} + g\underline{g} + b\underline{b}}{\sqrt{3}}$$

it is possible to establish a one-to-one correspondence between the four spin-flavour states and the four possible nucleotide pairs of DNA, the molecule which is depositary of hereditary information, and between the three colour states and the three distinct positions that a base pair can take within a codon.[6] Regardless of such a curious correspondence, a hard topic is to establish what happens to the pure states of the pion when the pion is fissioned into two pions by absorbing energy, since such pure states are unphysical. Obviously, each of the final pions is a quantum system in the same superposition of pure states as the initial pion, since they are neutral pions too, but in order to carry on this comparative study between pion physics and living organisms we should question whether it is legitimate to describe the process as a semi-conservative replication of each of the 12 pure states. A stance on this question, however, reasonably depends on the interpretation of quantum mechanics that is adopted. We therefore prefer to avoid such a digression in this preliminary work.

Regardless of our human limitation in understanding the superposition principle of quantum mechanics, once we have the description of the binary fission of a neutral pion in which it reproduces by absorbing free energy from the environment through a sort of semi-conservative replication of its quark structure, supported by a minimal anabolic pathway in which part of the incoming energy is converted into quark mass, we cannot avoid asking the crucial question: is a neutral pion alive? The answer can only depend on the definition of life you choose. In this regard, we simply observe that the definitions of life can be roughly grouped into three big families:

1. those based on complexity, metabolism, replication and reproduction;
2. those focused on the presence of a genetic program and on the inheritance of genetic information;
3. those focused on the fact that the system must undergo Darwinian evolution.

It is clear that the object we are dealing with is therefore to be considered alive in accordance with the first class of definitions,[7] while it would not be considered as such according to the last two. As concerns the second family of definitions, as we said, however, the description in terms of information and genetic program is a useful subject-relative description which tends to mask the reduction of biology to chemistry and therefore to the physics of inelastic collisions between molecules, ruled by the laws of electromagnetism. As for the last two families of definition, the neutral pion certainly does not appear to be a living system, but here we are really questioning whether the neo-Darwinian framework provides a complete understanding of what life is.

Of course the question does not concern only the neutral pion and its more massive versions $\eta, \eta', \ldots$, but also other mesons with different quantum numbers. For example, vector mesons, such as $\rho^0, \omega, \phi, J/\Psi, Y$, have the same quantum numbers as the photon $J^{PC} = 1^{--}$. Thus they are photoproduced in many experiments by photon scattering on nuclei, and they could, in principle, be 'photoreproduced' by photon scattering on vector mesons

$$\rho^0 + \gamma \rightarrow \rho^0 + \rho^0$$

in which a quark-antiquark pair would be 'photosynthesized' for the reproduction of the parent meson. Unfortunately, their mean lifetime is so short that this seems to be practically impossible; i.e. our universe doesn't seem to be fine-tuned for the thriving of such resonances.

Let's therefore continue to consider just the neutral pion. As has been said, neutral pions do not participate in atmospheric hadronic showers because of their short mean path length $\beta\gamma c\tau$ with respect to their mean free path $1/n\sigma$ in air, but also in liquid and solid substances; that's why we don't observe the proliferation of neutral pions in Earth's atmosphere. It is therefore a matter of verifying whether under some particular conditions these reproductive events can actually take place in the atmosphere, because this, as possible mental content in the author's mind, would constitute a good starting point for a possible semantic plane of understanding what life is. Clearly, neutral pion interactions with matter would be relevant if neutral pions crossed a much denser medium than Earth's atmosphere. In this regard it is useful to mention that in human-made experiments of high-energy ion collision, a pion gas is formed out of the cooling of a highly dense and hot medium which is produced in these events: the quark-gluon plasma (QGP). As we will shortly summarize, in the phase transition in which the hadron gas phase begins to appear, the interactions of neutral pions with the plasma are as relevant as those of charged pions given the high density of the medium. Of course, it is expected that the QGP is also produced in high-energy cosmic ray events, since the energy in the centre of mass can reach values comparable to those obtained in human-made ion collisions (LaHurd and Covault 2018).

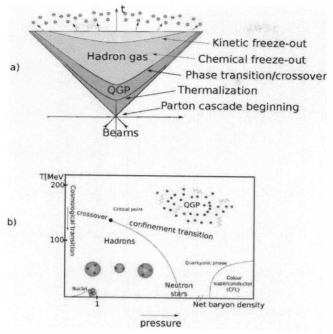

Figure 4.3 a) The temporal evolution of a fireball in heavy-ion collision; b) The QCD phase diagram: the cosmological QCD transition is believed to be a crossover like the EW transition. Such a diagram is often compared to the famous diagram of water in which, beyond the critical point, the vapour and the liquid phases are not clearly distinct any more

Following this idea, we therefore need to consider some QGP physics and specifically the interactions of neutral pions (appearing as a consequence of the cooling of the plasma) with the QGP out of which they appear. The production of QGP and the transition from QGP to hadron gas state is now an experimental fact. Experimental efforts at producing QGP under laboratory conditions started at CERN and BNL in the late 1980s (Baym 2016; Stock, n.d.). In the year 2000, after finishing the main part of its heavy-ion program at the SPS accelerator, CERN announced circumstantial evidence for the creation of a new state of matter in Pb + Pb collisions (Heinz and Jacob 2000). The actual discovery of QGP took place in 2005, when four international collaborations studying Au + Au collisions at the Relativistic Heavy Ion Collider at BNL announced the results of their first five years of measurements (Arsene et al. 2005). Surprisingly, the properties of the new state of matter differed markedly from predictions made over many years prior to its discovery. The description of the QGP as an ideal gas of asymptotically free quarks turned out to be only a rough description, whereas a detailed treatment requires accounting for long-range interactions through hydrodynamic models, in which the QGP is described as a droplet of a nearly inviscid fluid.

In such collisions, a fireball of QGP with an initial size comparable to that of colliding nuclei forms, and, as the fireball cools by radiating energy and expanding, the ordinary hadron gas phase begins to appear (fig. 4.3a). It is useful for our purposes to see how this state is predicted within the MIT bag model, in which the QGP is basically treated as a gas of free quarks and gluons, borrowing the tools of statistical mechanics typically used to describe a photon gas such as that contained in a black body, whose walls in this case are constituted by the bag. As is known, the pressure and the density of a photon gas depend only on the temperature, since in relativistic thermodynamics the number of particles is not conserved:

$$P\gamma = \frac{\epsilon_\gamma}{3} = \frac{\pi^2 k_B^4}{45\hbar^3 c^3} T^4$$

$$n_y = \frac{2\zeta(3) k k_B^3}{\pi^2 \hbar^3 c^3} T^3$$

Let's consider the expression of pressure P_π of a pion gas from statistical mechanics:

$$P_\pi = \frac{\epsilon_\pi}{3} = \left(g_B + \frac{7}{8} g_F \right) \frac{\pi^2 T^4}{90} = \frac{3}{90} \pi^2 T^4$$

where

$$g_B = 3, g_F = 0$$

are respectively the bosonic and fermionic degrees of freedom of a pion gas. The pressure of the QGP is instead given by

$$P_{QGP} = \frac{37}{90} \pi^2 T^4 - B^4$$

given that

$$g_B = 16, g_F = 24$$

for the QGP phase, and, considering that within the bag model the hadron bags merge into the bigger QGP bag, that gives rise to the term $-B^4$ of the bag pressure. For the Gibbs criterion the stable phase is the one with higher pressure, from which the critical temperature for the phase transition is obtained:

$$T_c = 150 \; MeV$$

In more refined calculations, such as lattice QCD, the estimation is about $170 MeV$, the so-called Hagedorn temperature, since Hagedorn first dealt with predicting the behaviour of nuclear matter at high energy, although his preliminary prediction was not correct. At such a temperature the quark density is given by

$$n_q = \frac{3 g_F \zeta(3)}{4\pi^2} T^3 \simeq \frac{1.35}{fm^3}$$

Although the QCD phase diagram is still being explored, much progress has been made since Cabibbo imagined a first pioneering phase diagram. The QCD phase diagram is illustrated in fig. 4.3b as it is currently known. As can be seen, the diagram presents a certain analogy with the famous phase diagram of water, in which the transition from the QGP phase and the hadron phase is currently believed to be a crossover until the critical point is reached.

QGP is therefore an exotically hot and dense medium produced in heavy-ion collision events and, reasonably, in high-energy cosmic ray events, even though it is harder to detect its appearance in the last case. These events in which the fireball appears are also called 'little bangs', as they mimic — according to the hot 'big bang' model of standard cosmology — the dynamics of primordial universe cooling: QGP is indeed the phase of nuclear matter in the quark epoch ($10^{-12}s - 10^{-6}s$). These little bang events reflect to a certain extent what happened in the early universe when the hadrons originated out of the cooling of the primordial soup (QGP) of the universe. In a qualitative description, the process is quite analogous to that of the formation of hydrogen atoms starting from e.m. plasma that occurs in the subsequent development of the universe at the time of about 380,000 years in the recombination era. Such a hadronization process is indeed called 'recombination' or 'coalescence' (fig. 4.4a). At the temperature of the phase transition, the quarks begin to form the first bound quark states and, in this way, the hadron gas phase begins to appear. But a peculiar process characterizes this transition with respect to a typical e.m. plasma-gas transition: as the hadronic phase begins to appear, the confinement mechanism begins to play a crucial role in the appearance of hadrons. Given that Hagedorn temperature is comparable to the pion mass, a collision between a pion and a quark of the plasma can give rise to the fission of the meson into subsystems, and, specifically, if the pion is neutral, its reproduction by binary fission has a certain chance of occurring (fig. 4.4b). Moreover, many body processes in which a quark of a daughter pion is taken from the plasma are not excluded due to the high quark density (fig. 4.4c); the average distance between the plasma quarks is in fact comparable to the size of the pion (1 fm). These processes continue to take place until their chemical freeze-out.

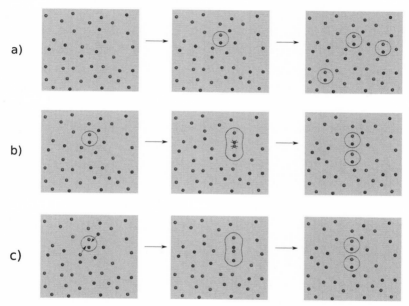

Figure 4.4 a) Coalescence; b) Meson fission by interaction with a plasma particle; c) Many-body interaction in which quarks of the daughter mesons are taken from the plasma

That is, pions are formed by coalescence, while the process responsible for possible pion reproduction is given by the interaction between the two phases. The pions receive energy for their reproduction from the same primordial soup of the universe out of which they are formed. So, in relation to the question "Is the neutral pion alive?" we find it interesting that such objects appear from the cooling of the primordial soup of the universe and can reproduce in such an environment through elementary anabolic pathways and semi-conservative replication, and, moreover, that these conditions are created again in the terrestrial atmosphere (as in any other celestial target) by virtue of certain high-energy cosmic ray collisions. We are focusing only on the most salient aspects that these objects have in common with living organisms. In a more in-depth analysis, it should be considered that the confinement transition in which these objects appear seems to be closely connected to the spontaneous breaking of chiral symmetry, although it is not yet clear whether chiral transition, i.e. the transition in which chiral symmetry is broken-restored, and confinement-deconfinement transition are the same transition. By resuming the formal analogy mentioned in Chapter 1 between phase transitions and dissipative structures, it is interesting to note that the emergence of these objects that present the basic characteristics of living organisms is therefore related to the spontaneous breaking of chiral symmetry within the strong interaction, whereas the emergence of living systems on Earth is certainly related to a chiral symmetry-breaking introduced by homochirality of organic molecules.

This comparative analysis between the physics of these objects and that of living organisms is interesting for us, because if it is legitimate to think at least of the neutral pion as a living organism, it is particularly significant that these systems are among the first ordered structures that appear in the thermodynamic path of the universe towards its thermal death. This, as we will discuss further, leads us to reflect on the possibility that—regardless of the inestimable progress that has been made since Schrödinger gave his famous lectures on the theme "What is life?"—we may not yet have fully understood what life is at the moment, and an approach in which life is described for what it appears to be, that is, a product of mind, could lead to some progress on the matter, showing a possible semantic plane of understanding of the phenomenon, thus connecting the problem of the nature of life to the problem we are interested in, namely the mind-matter problem.

We cannot avoid taking into account the fact that the semi-conservative replication of the quark structure of the neutral pion, and its one-to-one correspondence with the possible arrangements of nucleotide pairs in DNA linear polymers, pushes us to investigate the possibility that composite linear structures constituted of neutral pions could arise out of the cooling of the QGP. In practice it is a matter of exploring the possibility that DNA replication is the result of a mental reworking of an already existing process known to our supposed designer. If this were the case, it would mean that life can admit a semantic level of analysis; that is, it can be understood in reference to what it represents, as for example an orrery, a mechanical model of the solar system, is understood in reference to the solar system that it represents, and consequently one of our aims would be to explore what life on Earth could represent. We must therefore further investigate the confinement transition to verify whether the emergence of such linear structures of neutral pions is possible. To this purpose we consider that—as is well known—QCD becomes non-perturbative at low energy because of colour confinement. However, the approximate chiral symmetry present in the QCD Lagrangian and its spontaneous breaking allows one to construct a low-energy effective theory, with hadrons replacing the quarks and gluons as degrees of freedom. Chiral models, such as the linear sigma model (LSM), have long been used in many applications aimed at understanding various aspects of QCD, among them the description of disoriented chiral condensates in heavy-ion collisions or the chiral phase transition. We begin to consider its most simple realization, containing a triplet of pseudoscalar pions $\vec{\pi}$ and a scalar sigma field σ. In terms of the field

$$\Phi = \sigma \frac{\tau^0}{2} + i\vec{\pi} \cdot \frac{\vec{\tau}}{2}$$

it is easy to check that the Lagrangian is invariant under *SU(2)* chiral transformations, where the ϕ field transforms in the $(1/2, 1/2)$ representation according to

$$\phi \rightarrow exp$$

During spontaneous chiral symmetry-breaking, the σ field takes on a non-vanishing vacuum expectation value, which breaks *SU* down to *SU*. This results in a massive sigma particle σ and three massless Goldstone bosons $\vec{\pi}$.

We should therefore extremize the energy functional of the Lagrangian density to look for a solution in which the charged pion field vanishes, but fortunately this work has already been independently done. Indeed, after Vachaspati (Vachaspati 1992) showed that string-like structures exist in the standard model of electroweak theory, Zhang et al. (Zhang, Huang, and Brandenberger 1998) considered strings in models with spontaneously broken chiral symmetry of QCD and constructed a classical solution, the pion string, in the linear sigma model, as a static configuration of the Lagrangian in terms of the so-redefined fields

$$\phi \equiv \frac{\sigma + i\pi^0}{\sqrt{2}} = \frac{f_\pi}{\sqrt{2}} \left[1 - exp(-\mu r)\right] exp(i\theta) \quad \pi^\pm \equiv \frac{\pi^1 \pm i\pi^2}{\sqrt{2}} = 0$$

where the coordinates r and θ are polar coordinates in $x-y$ plane (the string is assumed to lie along the z axis). The string is as usual either an infinite length string or a closed loop string.

Then, even though the knowledge of this object is currently very poor, we consider that it can be roughly thought of as a linear object made out of π^0 and σ mesons; and, given that

Hagedorn temperature is comparable to the pion mass, while σ meson is more massive, we can approximately regard the string as being made only of neutral pions. Nevertheless, it would be risky to conclude that we have found the object we were looking for, because, as underlined by the authors, the string is not topologically stable, since any field configuration can be continuously deformed to the vacuum.[8] The formation of defects of various kinds and dimensions often occurs in phase transitions (Vilenkin and Shellard 2000), and, in quantum field theory, topological defects arise as solutions to partial differential equations that are homotopically distinct in the space of the order parameter with respect to the vacuum solution, and they are intriguing because they are permanently stable. Typical topological defects are, for example, vortices in rotating superfluid (Carlson 1996); topological defects in cosmology also constitute one of the possible candidates for solving the problem of dark matter. As is known, by constraining a subset of fields to vanish, if the vacuum manifold of the remaining unconstrained part of the system results in having a non-trivial homotopy group, then topological defect formation can occur, with this defect then being embedded in the larger theory (Zhang, Huang, and Brandenberger 1998), constituting the so-called embedded defect. However, the point is that, by the very nature of such embedded defects, their stability is not straightforward and needs careful analysis, since their existence is not strictly due to the topology of the full theory and they are usually not stable in vacuum. Indeed, similarly to the Z string in the standard electroweak model of the e.w. transition, the pion string is not topologically stable, since any field configuration can be continuously deformed to the trivial vacuum in the QCD sigma model, because the field values can escape in the constrained directions. With finite temperature plasma, however, Nagasawa and Brandenberger (Nagasawa and Brandenberger

1999) argued that the pion string can be stabilized. Basically, they replace the basic LSM with a linear sigma model with quarks included in the Lagrangian (LSMq) and study the stabilization effect due to the interaction with quarks. The idea developed by these authors is that the interaction with the plasma modifies the potential of the charged pions, giving rise to a subvariety in which the string is stable; therefore in the same plasma in which the string is produced, this would remain stable in a narrow range of temperature around the Hagedorn temperature of about 170MeV. More recent works by Karouby confirm the stabilization effect of this mechanism (Karouby 2013), as does the work conducted by Berera et al. (Berera et al. 2016). In these studies, it results that the string stability is dependent on the order of the chiral transition, and the string is unstable for a first-order transition. Thus, in order to have stability, the transition must occur above the critical point, where the transition is currently believed to be a crossover. Unfortunately, defect formation during crossover is very poorly understood both in analytical studies and in numerical lattice simulations, as the Kibble-Zurek mechanism is only valid for phase transitions. As a result, there is currently no well-founded model for following the evolution of strings. Huang et al. (Huang et al. 2005) also studied the possible signals of a pion string associated with the QCD chiral phase transition in LHC Pb-Pb collision, but as far as we know such an object has not been effectively searched for in available experimental data, and dedicated experiments have not been performed; so, it is currently predicted, but not yet observed or searched.

It is certainly honest to admit that at the moment we have no certainty about the stability of this object in the environment in which it appears, but on the other hand it must be considered that this does not necessarily constitute a limitation in this investigation. In fact, for our purposes, we are not directly interested in the fact that these objects actually exist in nature or that their existence is comparable to that of a pencil balanced on its tip, but rather we are interested in the fact that these may appear as mental contents in a mind involved in particle physics, and, as we have seen, they certainly appeared in at least some human minds interested in this field, for reasons completely independent from ours.

Anyway, if we assume that the strings—by interacting with plasma quarks—acquire some stability, it is interesting at least to get an idea of how these objects could evolve in the same hot primordial soup out of which they arise. The string is obviously subject to Brownian motion, but since the Hagedorn temperature is comparable with the mass of the pion, basically the string is not chemically freezed out in the environment in which it is stable, so we expect events of fission of the pions of the string. String-like defects, like others, are often subject to splitting and merging of topological defects in network configuration, so it is plausible that the child pions still belong to the topological defect; in this case they must be subjected to the constraint $\pi^{\pm} = 0$. This can induce replication of segments of the string, and, as we have seen for the scission of a single pion in QGP, it is possible that quarks of the plasma participate in the process of replication, just as available nucleotides are required for DNA replication, as explained earlier. The actual evolution of the string in its environment is a hard topic which would require further dedicated study, and we can just note here that the child segments have the same probability of replicating themselves as the original segments; that is, this object lacks any mechanism of inhibition of further replication, a mechanism which, according to neo-Darwinism, must have arisen at some point in the evolution of biological replicators (fig. 4.5).

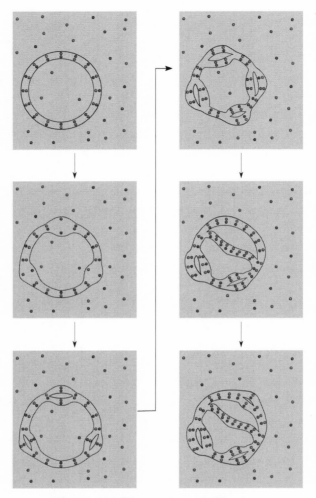

Figure 4.5 Pion string replication

Furthermore, in addition to the interaction between strings and plasma, it is useful to consider the interaction between strings. As far as we know, strings can, with a certain probability, intercommute, and, as closed strings, the intercommutation results in their joining, resembling the transposition process responsible for horizontal gene transfer that we mentioned in Chapter 1 (fig. 4.6).

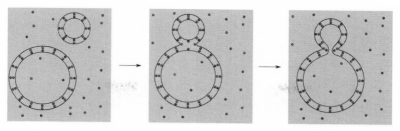

Figure 4.6 Intercommuting strings

The fact that these possibly existing linear objects, appearing as a consequence of chiral symmetry breaking, consist of neutral pions—which present the fundamental properties of living systems and whose wave function admits a biunivocal correspondence with DNA monomers—and the fact that the patterns of interaction of the strings with the plasma and between themselves are the same as those of semi-conservative replication and of horizontal genetic transfer, point to the possibility that the first living organisms may actually constitute some form of reworking by the hypothesized designer of these objects, which may constitute mental contents in the author's mind. In other words, there might actually be a way of thinking about the emergence of life on Earth as the production by the Earth itself—thought of as a subject of experience—of a meaningful response to the kind of stimuli we have discussed so far.

Clearly, the problem is that we have no certainty about the process of the emergence of life on Earth; therefore—in the terms we have proposed—we do not have a clear knowledge of the potential response of the alleged designer to the variety of stimuli they would have received from their environment. Consequently, we must follow a different approach, which, by the way, seems to be more in line with the current neo-Darwinian approach to the origin of life, in which plausible stories of physical events underlying its origin are presented, such as the one first proposed by Woese, for whom life would have originated among cloud water droplets. Having identified the physical events that act on the atmospheric system, and having discussed that it is possible that a perception of these cosmic ray events might take place in the author's mind, in accordance with the typical approach to the problem, we will rewrite the story of the origin of life presented in the previous chapter from a different narrative perspective: the author's first-person perspective. We will therefore exhibit, rather than a sequence of plausible physical events in third-person perspective, a sequence of plausible mental events in first-person perspective, which, starting from the knowledge of cosmic ray physics, lead to the emergence of a system recognizable as a plausible primordial form of life. As with the corresponding third-person story, we can't claim this first-person rewriting to be true. Its sole purpose is to highlight the fact that, just as typical neo-Darwinian stories should show the plausibility of abiogenesis, it is equally plausible that life has a semantic level of analysis. However, we obviously cannot state that we are certain of this, nor that we have identified its meaning, just as it is not legitimate to claim with certainty that life has no meaning since it is certainly not the product of a mind.

We will therefore now present such a plausible sequence of mental events—which we have attempted to infer—in the author's mind, whose 'brain'—the atmospheric electromagnetic system introduced in Chapter 3—must be responsible for the production of the first 'electromagnetic forms of life' on Earth, out of the stimuli—'nuclear forms of life'—we have seen in this chapter. Roughly speaking, given the mind-brain correspondence, since life might seem to be the result of a project from the mind of an author, once the possible 'brain' of an author has been identified and once we have examined the stimuli which such an author is subjected to, we just have to attempt to infer a possible project underlying the emergence of life on Earth. However, for literary simplicity, we will put ourselves in the clothes of the author; i.e. in the following chapter we illustrate the project as if we were the author, avoiding of course making any reference to living organisms as we know them, since in this mental setting they don't yet exist. In other words, we present a plausible so-called 'stream of consciousness', which must reflect to a certain extent what would have happened in the author's mind, by using the first-person narrative as if we were the author, even if, of course, we are not.

Chapter 5

A Living Organism as the Product of a Mind

As previously discussed, in every particle shower, as long as there is enough available energy, a multitude of matter particles belonging to the same class are produced (charged leptons in e.m. showers and hadrons in hadronic showers). In this process, whenever, for example, an electron-positron pair is created, a certain amount of energy ($2m_e = 1.022 MeV$ in this case) is converted into mass. The point that captures our attention is that the amount of energy that can be converted into mass in each interaction is a discrete quantity, accurately detailed in the periodic table of elementary particles, in which particles are characterized by mass, spin and charge. We do not know the reason—if it makes sense to speak of a reason—for this basic discreteness of nature;[1] i.e. we don't know why matter can be reduced to a handful of elementary particles. We only know that elementary particles are characterized by discrete physical quantities, unlike, for example, a dust particle, which instead has approximately continuous mass, angular momentum, and charge. Such basic discreteness of nature is the reason why in a cosmic ray shower more identical physical objects, i.e. physical objects with the same values of their characteristic physical quantity, are produced and produced again several times; that is, they are iteratively reproduced. Thus, the proliferation of identical physical objects in a shower has a certain appeal for us, since it is a consequence of a peculiar characteristic of the world of ultimate elements of nature; as we said, it does not so evidently apply to a particle of dust. And certainly the most interesting case is that of hadrons, because, even though they are not elementary particles—rather they are composite objects, i.e. particle systems—a multitude of identical hadrons is produced in a hadronic shower. More specifically, as they are systems rather than elementary objects, it makes sense to speak of an energy flow between the system and its environment, and, in the previous chapter, we briefly described the process by which hadrons reproduce themselves by virtue of quark confinement, i.e. more hadrons are produced by absorbing energy from their environment. In other words, these close and mysterious discrete constraints on particles that can exist give rise to the phenomenon of reproduction of particles, or equivalently to the growth of a population of particles in a shower. Hadrons seem to be particularly interesting in this sense, since they are particle systems rather than elementary particles, but the same characteristics of reproduction and growth observed in e.m. showers—made out of elementary particles such as leptons and photons—are observed in hadron showers—made out of quark systems. Furthermore, since there are no stable mesons—which are the main constituents of hadronic showers—the growth of the population of particles in a shower interestingly concurs with its decay, as if they needed to absorb energy to maintain themselves and thrive; otherwise their existence ends by decay, ultimately into leptons and photons. We therefore are generally interested in studying the behaviour of quark systems under energy absorption, since we want to comprehend the extreme consequences of such a charming discreteness of nature in terms of reproduction and growth.

Among these particles, the neutral pion is the most interesting, since, in principle, exactly identical objects can be produced through energy absorption, leading to the growth of the number of such particles. Its appeal for us also lies in the fact that its physical state is due to a superposition of physical, observable eigenstates such as flavour and spin, and we certainly struggle a lot to really understand what this truly means. However, as previously discussed, its lifetime is so short that such reproduction events can only happen in an exotically dense medium such as the QGP—the primordial soup of the universe out of whose cooling they form—which can be recreated in human-made experiments, as well as in high-energy cosmic ray events. Let us start, then, by just comparing these two types of above-mentioned nuclear objects: a neutral pion and a QGP droplet. In a neutral pion—as in any hadron—there is an equilibrium condition between the kinetic energy of the quarks and the potential energy of the bag, while, when the bags merge in forming the QGP bag, the vacuum potential is not enough to balance the hydrodynamic pressure and the fireball expands typically according to linear law in time.

At this point, we naturally become interested, at least as a theoretical study, in seeing how these fireballs would behave in hydrostatic equilibrium rather than in the typical expansion observed; that is, we wonder how such a fireball would behave under the effect of an external pressure that establishes hydrostatic equilibrium, as happens for individual hadrons, since we suspect that such a condition of hydrostatic equilibrium could be crucial for reproduction and growth, disregarding for now the question about the actual feasibility of this system.[2]

Let us therefore suppose that we apply to a fireball, at a temperature slightly above Hagedorn temperature with null baryon chemical potential ($\mu_B = 0$), an inward pressure

$$P = \frac{\epsilon_{QGP}}{3} - B^4 = \frac{37}{90}\pi^2 T^4 - B^4$$

which exactly balances the outward pressure given by partons, which are assumed to be free in the bag model. The fireball in such conditions would lose energy only by radiating, and the simplest model to treat it would be the black-body model. However, typical human-made fireballs do not exceed 25fm in size, and cosmic ray fireballs are expected to be of comparable size, given that they would be produced at a comparable centre of mass energy. The crucial point here is that, despite their high density, they are quite transparent to radiation because of their small size, so that radiation occurs from the volume rather than from the surface, and consequently the black-body radiation model is not valid for them. Similarly, because of their transparency, all energy absorption from the environment can be overlooked. We must therefore consider, in this theoretical analysis, the smallest opaque fireball, which is about $10 pm$ in size (Martinez 2013). That is, we are considering the smallest fireball that truly shows the characteristic opacity of primordial QGP, but in different thermodynamic conditions: basically it is a non-isolated system (radiating and absorbing) since it lives in an already existing universe and is somehow at constant pressure. It therefore shows an emission peak of γ-rays with energy given by Wien's displacement law:

$$E^{peak} = h\nu^{peak} = \alpha k_B T \simeq 479 MeV$$

At constant pressure and therefore constant energy density

$$\epsilon = \frac{37\pi^2 k_B^4}{30\hbar^3 c^3} T^4$$

the system must decrease its volume due to the effect of radiation:

$$\epsilon \, 4\pi r^2 dr = dU = dQ - PdV = -\sigma T^4 4\pi r^2 dt - \frac{\epsilon}{3} 4\pi r^2 dr \, \frac{4}{3}\epsilon \, dr = -\sigma T^4 dt \frac{dr}{dt} = \frac{-3}{296} cr$$

$$= r_0 - \frac{3}{296} ct$$

More generally, if a net energy flux ϕ of γ-rays between the system and its environment is considered, by defining ϕ as positive in absorption and negative in emission, and expressing it as $\phi = k\sigma T^4$, in terms of the black-body radiant emittance and a multiplicative factor k, then one gets

$$r = r_0 + \frac{3}{296} kct$$

Thus, in the absorption regime its volume grows according to cubic law in time, certainly keeping constant its quark density, just as is the case for a photon gas at constant pressure, whose thermal capacity is formally infinite and which undergoes isothermal expansion by absorbing energy from an external thermal bath present in its environment. Similarly to the photon gas, the free energy dissipation involved in the process is quantified by the entropy increment

$$dS = \frac{dQ}{T} = k\sigma T^3 4\pi r^2 dt = \frac{296}{3}\frac{\sigma}{c}T^{34}\pi r^2 dr = \frac{74}{45}\frac{\pi^2 k_B^4}{\hbar^3 c^3}T^3 dV$$

However, what is different with respect to the photon gas at constant pressure is that here we are dealing with quarks, which are massive particles; therefore it must be taken into account that a fraction of the incoming energy is converted into the mass of the quarks, and, conversely, if the net flow of energy is outgoing, there is a transformation of mass into energy. The ratio $\frac{\Delta m}{\Delta Q}$ of absorbed energy which is converted into mass can be obtained as follows:

$$\Delta Q = \frac{4}{3}\epsilon \Delta V \Delta m = n_q m_q \Delta V \quad \frac{\Delta m}{\Delta Q} = \frac{3}{4}\frac{n_q}{\epsilon} m_q \simeq 0.5\%$$

where $m_q \simeq 3.5 MeV$ is the average mass of u, d quarks.

But regardless of the fine calculations concerning this relativistic system, which of course would require more refined hydrostatic models, the crucial difference between this system and a photon gas at constant pressure is that, as we mentioned, even though we are able to create a constant-pressure photon gas that exchanges energy with the environment—since we can make it not so hot and not so small—we have no idea of how to create the described system. It therefore seems to be basically unphysical compared to a typical expanding fireball, because of its size ($10pm$) and its condition (hydrostatic equilibrium). That is, even though this system has appeared among our mental contents, it is at the moment a mental content which no physical objects correspond to. Such a condition is a typical condition of the first steps of any creative process, although we have no idea of how to proceed and reach a physical realization of this mental content.

It seems then that our exploration should stop here. However much we manage to get an idea of how something like this could evolve, we have no idea of how it can be realized, if indeed it were actually possible to realize it. Disregarding the desire to somehow realize it, we can only stop and contemplate these high-energy collisions, in which 'a little expanding universe' seems to occasionally appear in Earth's atmosphere. What now catches our attention is that, among all the cosmic rays that interact with Earth's atmosphere, certainly, when a cosmic particle crosses a cloud, it has a greater probability of interacting with matter, given the greater density of the liquid state compared to the gaseous one.

Our attention therefore shifts to cloud water droplets. As discussed in Chapter 3, liquid water can condense around a smaller droplet surrounded by an amphiphilic double-layer membrane. Let us thus suppose the existence of a droplet containing a water-filled double-layer membrane of about $20\mu m$ in diameter, in which different compounds are dissolved; in particular, we happen to find that, within such a membrane, approximately equal numbers of four distinct molecules belonging to the same class of compounds are confined, since the membrane happens to be impermeable to them. Suddenly, a high-energy cosmic ray that has passed through most of the atmosphere impacts one of the nuclei of matter within the region defined by the membrane, resulting in a 'little bang' event.[3] Now, the image of the system that we previously abandoned—the constant-pressure fireball—resurfaces in our mind, since what we are looking at somehow reminds us of it: these four molecules confined within a membrane seem like free quarks in a bag (fig. 5.1).

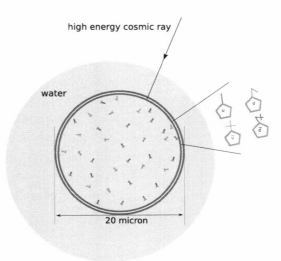

Figure 5.1 Water droplet in which four distinct compounds representing quarks are dissolved and which is separated from an aqueous environment by an amphiphilic double-layer membrane

Immediately an idea arises in our mind. In this case, if we do not intend to abort the creative process we are interested in, the best way we can think to proceed is to try creating a physical model of the system that we have outlined, a material representation in more comfortable conditions of pressure and temperature, that is, a physical system that for us in certain respects stands for the physical system outlined, similarly to how an orrery (a physical model of the solar system) is a physical representation of the solar system. The production of physical representations is typical of the most advanced activities of the mind such as art and science. Like an orrery, the Crick model of DNA is a famous example of representation, but, unlike in the construction of an orrery, or of a physical model of DNA, here we are involved in the construction of a representation of a physical system which we don't believe exists yet, and neither do we know how to build it. We just want to see our idea somehow realized; we are therefore involved, in this sense, in a creative process. Like any representation, whatever we manage to achieve, it will never fully reflect all the aspects of the system it represents (otherwise the distinction between the object and its representation would disappear); for example, Crick's physical model of DNA does not in any way represent DNA Brownian motion (unless someone starts to randomly beat the model, of course). We therefore want to represent the system discussed above with a non-relativistic thermodynamic system, an ordinary chemical system that we expect we would be able to realize, which is for us a representation of the relativistic system we have explored: a chemical system in which certain molecules for us stand for quarks. Asking the reason or the purpose for which we are doing this is like asking the reason or the purpose for which humans have been involved in art and science since the dawn of humanity. But in this case we could answer that, as we have said, given the intrinsic property of hadrons in terms of reproduction and growth, we are interested in figuring out the behaviour of a more complex, asymptotically free quark system under the same equilibrium configuration as hadrons, which is basically a little hot universe at constant pressure. It is a legitimate study from a scientific point of view, and it is an amazingly creative artistic process too.

We therefore choose those four distinct molecules we found within a 'chemical bag' belonging to the same class of chemical compounds, to represent the four distinct quarks, which are roughly assumed to be free, and we will refer to such molecules as 'quark molecules', i.e. they are the molecules that for us — the designers of a chemical system which must mimic the behaviour of the above relativistic system — stand for quarks in the system to be represented. From now on, we will refer respectively to the two systems as 'the real system' and 'the representative system', or briefly the 'representation', for literary simplicity and of course to underline the relation between reality and its representation.

We must therefore briefly discuss the scale factor for the representation we intend to make. Given that the system we wish to represent is about $10pm$ in size, and the coated water droplet we have considered is about $20\mu m$, the space scale factor is then about $2 \cdot 10^6$, and consequently, if the quark density is assumed to be of the order of $1/fm^3$, then the molar concentration of each of our quark molecules will be $1/(2 \cdot 10^6 \ fm)^3 \simeq 0.2mol/l$. In this representation, the original system is therefore magnified and, of course, represented in slow motion, since quark velocity is near the speed of light while the root mean square speed of molecules with mass, e.g. m=500Da, is about $v_{rms} = \sqrt{3K_B T/m} = 125m/s$.

At this point, we want our molecular representation to mimic the behaviour of the real system, the constant-pressure fireball. Specifically, we want the system to increase in size by absorbing energy from its environment while keeping the concentration of quark molecules constant, just as the real system grows by keeping the quark density constant. The problem is that obviously our molecules are not elementary particles and therefore it is not possible to produce them in pairs by transforming energy into mass; we must therefore find some trick in order to proceed in our intent. As explained earlier, the spherical double-layer membrane distinguishes an aqueous system with respect to its aqueous environment, and it is selectively permeable, in the sense that, even though it confines quark molecules within its inside region, it allows simpler compounds to pass through it, and such a property can be to our advantage for our purposes. Indeed, if the system is placed in an environment where simple compounds are present, a mass flow directed towards the inside of the system will tend to balance the concentrations (internal and external) of these compounds. Let us suppose that among these simple compounds there are all the precursors necessary for the chemical synthesis of the quark molecules. Given that they must be complex molecules, their synthesis typically requires an energy input in order to break highly stable bonds of simple compounds and rearrange their basic constituents. In practice it is a matter of realizing the synthesis of the molecules within the system, using the free energy that it absorbs from its environment as incoming radiation. Once synthesized, these complex molecules will remain confined within the system, like newly created quarks in the bag, and the creation of quark molecules will effectively require the conversion of energy into mass, even though the amount of converted energy will be orders of magnitude lower for molecules with respect to quarks. As previously discussed, in the case of quarks $\frac{\Delta m}{\Delta Q} \simeq 0.5\%$, while in the case of molecules (whose binding energy is of the order of eV) this ratio will be of the order of $\frac{eV}{100 Da} \simeq \frac{eV}{100 GeV} = 10^{-11}$. With an appropriate tuning between the mass flow through the system and the energy flow, the internal and external concentrations of these simple compounds will remain at a certain distance from equilibrium, since within the system they are reactants for synthesis of quark molecules. The great majority of the synthesis reactions give rise to waste products that can flow through the membrane, tending to balance the internal and external concentrations of the reaction waste. However, this expedient would lead to an increase in the number of quark molecules, but if we want the concentration of these molecules to remain constant like that of the quarks in the constant-pressure QGP, we must ensure that the representative system grows adequately in size. The system must therefore also be autonomous in the synthesis of membrane amphiphilic molecules, thus increasing in volume too, and keeping the concentration of the quark molecules constant (fig. 5.2a). Clearly, in the representative system we are designing, the free energy will not be constituted by gamma rays; but visible solar photons of 2-3 eV, or at most ultraviolet photons of about 7eV, will be fine for initiating the reactions. Obviously, the chosen expedient can be used to represent the fireball in a radiative regime too. In the absence of an external energy source—let us say during the night—the quark molecules (and membrane molecules too) must undergo a complex of decomposition reactions that will decrease the number of quark molecules by keeping their concentration constant, and the mass defect of these reactions will contribute to the emitted energy. With an appropriate fine-tuning of the reaction rates, the system will therefore reduce in size, keeping the concentration of quark molecules constant, emitting radiation (fig. 5.2b), as well as obviously releasing the reaction wastes into the environment.

At this point, some overall considerations on the feasibility of our project are necessary. Let us remember that we have chosen to start this work of representation because we have no idea of how to create the real system that interested us, but in proceeding in the work of representation it already seems clear enough that what we intend to achieve is in fact a hugely difficult task for the most skilled chemist. In practice, nobody at the moment knows how to achieve what we are talking about, but nevertheless we can at least get an idea of how we could try to create such a chemical system, while in the case of the constant-pressure fireball we have no idea of how to do it, assuming it can be done. More specifically, if the absorbed energy must initiate such a complex of reactions, we could try to temporarily store the energy in some high-energy bonds of a specific molecule or of a small number of such molecules, which we will use as energy carriers, i.e. carriers of the energy that feeds otherwise unfavourable reactions, which require the breaking of such high-energy bonds. Certainly, among the most suitable molecules there are those that have a chain consisting of two or more identical molecular ions, since they are formed by high-energy covalent bonds because of their intramolecular electrical repulsion. Of course, to avoid setting up a specific synthesis complex for such an energy vector, the simplest possibility is that such an energy vector must be a molecule already involved in the transformative complex we wish to set up. Assuming we chose, e.g., a phosphate ion PO_4^{3-} to constitute the molecular chain, we could figure out the following scenario: our energy vector will be a precursor of the synthesis of a quark molecule and will exist in the form in which it contains a single phosphate ion or a chain of two, three, or possibly more phosphates: preMP, preDP, preTP, depending on the energy it carries.[4]

absorbing   radiating

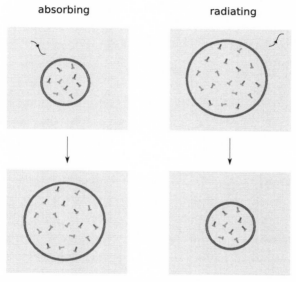

Figure 5.2 a) Growth of the system in absorbing regime; b) Decrease of the system in radiative regime

Assuming therefore that, with such expedients, it is possible to realize such a system, the question we are asking now is this: how faithfully does this chemical system mimic the behaviour of the system it must represent? More specifically, in line of principle the radius of the real system can grow linearly indefinitely over time. For this to occur, the condition on the radiative flow $\phi > 0$ must be satisfied, or, considering the more general case in which the flow is a periodic function with period T, the condition is $\int_0^T \phi(t)dt > 0$. It is therefore necessary to consider the case in which even the average energy balance of our representation over time is positive. This can be achieved with a system that absorbs visible photons during the day and emits infrared photons during the night. However, we will not go into the details of the system's energy balance — in which of course non-radiative heat processes must be considered too — since the realization of a positive energetic balance is not expected to constitute a problem. The crucial point of the question is that the system, as we have thought of it so far, even under conditions of appropriate energy balance, cannot grow indefinitely for various reasons. Let us start by looking at one. The problem is that for most reactions that must occur within the system, specific catalysers or, more generally, specific molecular devices are necessary; e.g. they must couple the hydrolysis of the energy carrier with the unfavourable reactions. However, as the system grows, the concentration of these molecular devices certainly decreases, consequently lowering the chemical kinetics of these reactions, ultimately to a value no longer sustainable by the system. The only solution we can think of is that, just as the system autonomously synthesizes quark molecules and membrane molecules, it must autonomously synthesize all the necessary molecular devices responsible for such transformative processes. It is therefore necessary to implement a complex of reactions responsible for the production of these accessory molecular devices. On the other hand, if we tried to proceed in this way we would soon realize that, given the quantity of molecules to synthesize and their complex structure, it becomes impossible to set up a dedicated synthesis pathway for each of them. This is not the only problem that an indefinite growth of this system presents, but since this does not seem to be of easy solution, we choose to face them one at a time and therefore we begin by focusing our attention on this one.

It is useful, in this regard, to analyse similar cases in which a representation of a physical system is built, in order to find a way to proceed. Let us, then, go back to the orrery we mentioned as an example of representation, a physical model which represents the behaviour of the solar system using gears. If the system quantitatively reflects the motion of the solar system bodies, it is in fact an analogue computer that computes the orbital motions. The most ancient analogue computer that predicted astronomical positions and eclipses for calendar and astrological purposes was the Greek hand-powered computer known as the Antikythera mechanism. But today of course we typically don't play with the gears of an analogue computer when we wish to represent the motion of the bodies of the solar system; rather, we typically prefer to use a digital computer that writes on a data storage medium, such as tape, the data encoding the information that defines the orbits, and of course the system must be endowed with a mechanism responsible for decoding the information so as to direct the gears, coupled with the degrees of freedom of the spheres we use to represent the planets. If the computer is a universal computer it must be programmed first, but the interesting point is that the two parts of the system can also be decoupled. In practice we could produce as much data about the orbits as we liked and store it on a data storage medium; then the piloting mechanism could independently read such data to pilot the spheres. Returning to our project of representation, we observe that we could equally try to resort to digital technology to overcome the aforementioned problem. Suppose that all the necessary molecular devices can be created by somehow arranging a finite number of simpler units which the system can synthesize autonomously.[5] In this case we should first somehow produce the data that encodes the composition of all the molecular devices the system must synthesize and then equip the system with a data storage medium, such as a molecular tape, which keeps the information for the synthesis of all the necessary molecular devices, including of course those machines that are dedicated to the synthesis of all the molecular devices by decoding the stored information; we will call them decoding machines (decMachines).[6]

But this gimmick alone would not be enough to solve our problem concerning indefinite growth, since, given that the density of molecular devices must be kept constant to have indefinite growth, even the density of decMachine-tape systems must be kept constant. As regards the growth in the number of decMachines, these can be synthesized like all other molecular devices by other decMachine-tape systems, but the point here is that, unless we let the molecular tapes somehow replicate, the density of decMachine-tape systems cannot be kept constant, because of lack of tapes. In other words, in order to keep the reaction rate of the decoding reaction constant, the data density within the system must be kept constant as well.

Let us reflect on this issue a bit more. In order to let the tape somehow replicate, the system should surely be autonomous for the synthesis of the basic units that constitute the tape, since simple compounds are generally assumed to be the only ones present in the system environment, and we need some specific molecule to compose the tape. In this regard, the simplest possibility is that the components of the tape must be molecules that the system already synthesizes autonomously, and of course we cannot fail to mention the intrinsic replication property of those quark systems we discussed in the previous chapter. So, the most elegant possibility we can think of would result in the following dilemma: the tape replication should mimic some physical process of the system we wish to represent, but unfortunately, at a temperature above the Hagedorn temperature—the condition we have chosen for the real system—no stable quark structures are present.

***

We therefore resort to the following choice: we will consider the real system at Hagedorn temperature, at which pion strings are supposed to be stable (fig. 5.3).

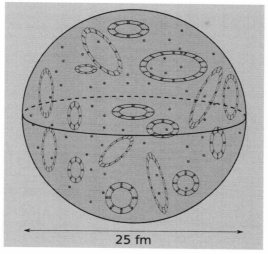

Figure 5.3 Pion strings within a fireball of QGP of typical size

We are therefore introducing a modification to the initial project. That is, we become interested in creating a physical model of a fireball in conditions of hydrostatic equilibrium that can exchange energy with the environment, but in the narrow range around Hagedorn temperature at which pion strings are expected to coexist with approximately free quarks. The new objects to be represented with molecules are in this case the pion strings. Since a pion is a superposition of pure states, the string itself will be a superposition of states, but it is not possible to represent superposition in the mode of representation we have chosen, because clearly each of the four quark molecules cannot be superposed with the others; thus, all we can do is represent a pure state of the string. If distinct quark and antiquark flavours are represented by distinct molecules, we can choose to represent antiparallel spins with opposite orientation of such molecules, which will thus constitute a double-stranded polymer. We will refer to such a polymer as a 'string molecule', since it represents a pion string of the real system we wish to represent, and, on the other hand, it also constitutes a useful data tape for our purposes. The simplest realization choice is of course that all upward molecules lie on a strand, while all downward molecules lie on the other strand, which results in a double-stranded antiparallel polymer (fig. 5.4). Furthermore, if we understood the relationship between confinement transition and chiral symmetry breaking, we could somehow represent the spontaneous breaking of chiral symmetry in our physical model made with molecules, but this seems to be premature given our current knowledge of the process.

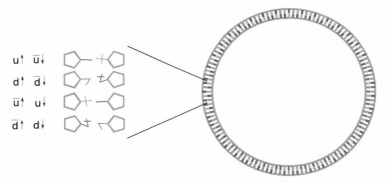

Figure 5.4 Molecular representation of a pure state of the pion string

At this point, we need to somehow encode the information for the synthesis of all the machines necessary for the system, which must be constituted by a finite number of simple units. In this regard, we observe that certain polymers undergo molecular folding in water and acquire a complex form which is basically dependent on their specific constitutive sequence, by virtue of the specific self-interaction between their parts and with surrounding water molecules. Such folded linear molecules can join together to form more complex structures by virtue of the residual intermolecular bonds. We therefore choose to use this property in order to simplify, as much as possible, the synthesis of all the molecular devices the system needs.

Having four distinct symbols to encode the composition of a molecular device made out of a sequence of n distinct possible building blocks, the minimum dimension of the data packet which encodes for a building block is given by the minimum number of symbols $m : m \geq log_4 n$, and unless n is a power of 4, encoding occurs with degeneracy, which is also useful with respect to the management of possible data losses on the tape, as well as the redundancy given by the fact that the tape is double-stranded.[7]

Having introduced strings into the system to be represented, we just have to represent the interaction of the strings with plasma and of the strings among themselves. Let us start by representing the interaction of the strings with plasma. As we have mentioned, since the Hagedorn temperature is comparable to the mass of the pion, it is plausible that in the interaction of the string with the plasma, segments of the string can replicate, giving rise to the phenomenon of splitting and merging of topological defects. The interactions can be rather complex in that plasma quarks could participate in the semi-conservative replication process of the pion quark structure. In the case of our chemical representation of the system, certainly the nucleotides necessary for the replication of the molecular tape cannot be created in pairs out of interaction energy with free-floating quark molecules; therefore the only possibility we have is that nucleotides already present in the system participate in the replication process, representing a plausible physical process of the real system. On the other hand, the mechanism of tape replication is what ensures the necessary data flow throughout the system, which is necessary, as we have seen, for its indefinite growth.

The realization of the system is subject to the following issue: our molecular representation of a pion string (in brief, a string molecule) is a polymer made out of quark molecules, which are present as free-floating quark molecules in the system. However, even though quark molecules can impact against the string molecule, such an impact does not give rise to any replication; this is not surprising because the physics of this system is completely different from that of the system we intend to represent. The process therefore requires a specific work to take place. In fact it requires an energy contribution, which will be provided by an energy carrier, such as the preTP, whose hydrolysis feeds a dedicated preTP-dependent machine or a dedicated machine complex for replication (fig. 5.5). These machines will then have to hook in random points onto the string molecule and slide on it, catalysing its replication.[8]

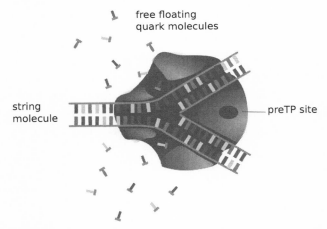

Figure 5.5 Tape replication process in the pool of free-floating quark molecules as a representation of the interaction of the pion string with the asymptotically free quarks

Certainly, there is a problem with this process: even if the DNA replication mechanism gives rise to multiple uncontrolled replications of segments, as is expected for a pion string, this is not as efficient in terms of data copying. As a result, we would be faced with the problem of constantly having multiple incomplete copies, which are unavailable for decoding. Hence, we must necessarily represent a physically possible event, as improbable as it is, namely, that the strings replicate exactly without ramifications.

Having accepted this implementation compromise in the representation of the interaction of strings with plasma, now we move on to the representation of string-string interactions, which, as topological defects, are expected to intercommute. But, as our molecular tapes are not topological defects, they will not spontaneously merge into a single polymer when interacting. We must therefore build a specific machine (whose composition must be encoded in a molecular tape) that attaches to two string molecules and catalyses their union, this way representing the strings' intercommutation (fig 5.6). It must be considered, however, that given that each tape encodes useful information, the machine cannot be attached at random points of the two molecules, but must recognize the specific sequences that encode the beginning or the end of a data packet. The interaction will therefore result in two data packets that encode information for building two molecular devices—or more than two if the process is reiterated—on the same tape.[9]

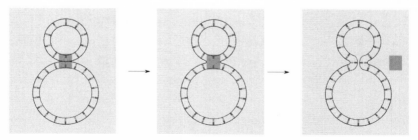

Figure 5.6 Molecular representation of intercommuting pion strings

Finally, while we have resorted to digital technology to solve a problem concerning indefinite growth, the system as designed so far is certainly not yet ready for indefinite growth to occur. It is indeed easy to verify that because of intrinsic limitations to the physics employed in this realization, such a behaviour is not possible: when a spherical system of radius r grows, its volume increases more rapidly (r^3) than its surface (r^2); so, the supply of nutrients and the expelling of wastes by diffusion or by virtue of dedicated devices become more and more inefficient as the system grows.[10] Therefore, the necessity of exchanging matter with the external environment introduces a lower limit to the surface/volume ratio of the system, and, considering the system as spherical, given that this ratio is inversely proportional to the radius of the system, an upper limit to the size is required of the system.[11]

We must, then, find a satisfactory way to proceed and overcome this problem. It is useful to return to the starting point of our idea, namely, the creation of a fireball that reflects the equilibrium inherent in a neutral pion (or more generally in a hadron). Let us just note that, when energy is supplied to a pion, it splits into two or more pions, giving rise to systems for which the surface/volume ratio is kept high. As regards the constant-pressure fireball to which energy is supplied, it—unlike the pion—does not necessarily split into smaller droplets, but this does not imply that the split is a prohibited physical event; as far as we know, nothing prevents the real system from splitting in two once a certain size is reached. On the other hand, the only means we have to let our representation grow indefinitely is to keep the total surface/volume of the system ratio high, and this would be certainly possible if the growing system gave rise to disconnected regions. In other words, what we can do is to represent a constant-pressure fireball that reflects the behaviour of the single pion when absorbing energy; that is, it splits into two or more systems when a certain size is reached. Although, as far as we know, the real system we are attempting to represent does not necessarily perform this split as a single meson does, the split is, in principle, a possible physical event and, on the other hand, necessary to our mode of representation. In this further variant of the initial project, the representative system grows indefinitely by increasing the number of disconnected regions; that is, each of the two child subsystems must give rise to fruitful offspring. In practice, in order to realize it, the replication mechanism must be finely regulated in line with the growth rate, while it is also necessary that there be a certain redundancy of the hereditary information, in order to ensure with a good probability that, at the time of the split, at least one copy of each initial tape necessary for the functioning of the system is in both child subsystems. Certainly, if such redundancy proves to be unsustainable for the system, a suitable partitioning system[12] must be provided to ensure the stable transmission of the molecular tapes on which data is stored (fig 5.7).

a)   b)   c)   d)

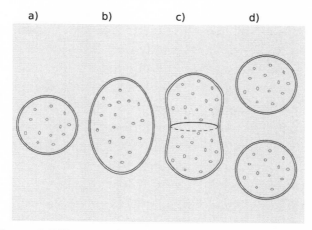

Figure 5.7 Fission of the system; string molecules are illustrated magnified and in small numbers

Let us also note that in this reformulation of the initial project, the total volume growth—which is proportional to the growth of the population of systems that reproduce the original one—is exponential rather than cubic as in the initial formulation.

Overall, it seems that—starting from a reflection on interesting characteristics of the physics of cosmic rays—we have come to a fairly stable formulation of our project. We wish to represent an exponentially growing population of little hot universes, at hydrostatic equilibrium, which absorb energy from their environment, in other terms, a 'shower' of little hot universes at constant pressure which reproduce themselves by absorbing energy from their environment. And, given the need to implement a data replication mechanism, we have chosen its temperature so that free-floating quarks can coexist with self-replicating pion strings, which we represent as replicating and merging molecular tapes.

Once such a project has taken a stable form in our mind, if we believe that such a work can be done, we should actually begin to gather some chemicals and do it. As we have mentioned, this is by no means trivial, basically because, even if we could manage to realize its single components, we don't know yet where to start to build the whole thermodynamic system. Once again—as for the real system—we find ourselves confronted with the great distance between our intentions and our actual realization capabilities. However, in this case, we at least have an idea of how to attempt to create such a representation: probably a good starting point would be working on the simplest replicating tape we can realize, surrounded by a pool of free-floating components, postponing the problem of how to pass from this possible preliminary system to what we finally intend to achieve. Another big problem concerns how to obtain the data necessary for the functioning of the system, and, in this regard, the most rudimentary way of proceeding that comes to mind is starting from a random set of data and seeing what happens. Anyway, such a complex creative process is certainly not a small undertaking, but we are confident that, with repeated efforts, something like this can actually be achieved.

Chapter 6

A Beautiful Question: Is Life a Work of Art?

In the previous chapter we presented a possible sequence of mental events that can take place in a mind interested in the physics of cosmic rays. Starting from a few characteristics peculiar to these physical events, we gradually formulated a design for representation, which can be placed on the border between art and science.

Of course, if we secretly managed to complete such a project and if the system we created were to be found somewhere in nature by oblivious discoverers, such a system would certainly be recognized as a form of life, given that—as a representation— it evidently shows the characteristics of the system it represents, namely, metabolism and replication. Specifically, if we used phospholipids as membrane molecules, dNMPs as quark molecules, ATP as the energy carrier (preTP), proteins as molecular devices or their parts, and the same genetic code as life on Earth, the system would be recognized as a primordial form of life—probably as an ancestor of photosynthetic bacteria, such as cyanobacteria—which, however, has its genetic material completely scattered on the plasmids and whose main energy reserve consists, rather than of glucose, of the dNMPs, i.e. of the ribose of which they are made.[1] In fact, all the difficulties we mentioned in the previous chapter are due to the fact that the technology required to make such a work is actually the technology necessary to make life from scratch, which nobody is known to have as of today.

The previous chapter constitutes, therefore, a rewriting from a first-person perspective of a typical story about the origin of life among lightning discharges and dirty water droplets, usually narrated from a third-person perspective, even though in such narratives questions concerning whether certain mesons can be considered living systems are generally overlooked, just as the causal power of cosmic ray events over lightning discharges is neglected, together with everything else that may possibly follow.

However, if we secretly managed to complete the project and if the system we created were to be found somewhere on Earth by oblivious discoverers, the discovery of our masterpiece would likely be interpreted as a confirmation of the neo-Darwinian conception of the biological world, representing the discovery of a missing link preceding the appearance of prokaryotes as we know them today, which are very much the same as they have always been. This link would be conceived as having appeared, like any other, by virtue of a 'blind' and 'unconscious' process, to use Dawkins' words. The interesting point to note is that, even though the imaginary oblivious discoverers would completely understand the physico-chemical details of the implementation, they absolutely would not understand what the object they have found is, i.e. a representation of a little hot universe at hydrostatic equilibrium which grows by absorbing energy from its environment, realized by taking some creative freedom so as to meet a few realization needs. However much they were able to analyse and describe the object in purely materialist terms, they absolutely would not understand what it was; that is, its meaning would remain totally hidden, precisely because in principle its meaning would not be sought. They would therefore be in a position similar to that of a person who described, for example, the Antikythera mechanism purely in terms of the physics of the gears, without grasping any reference to the solar system.

To sum up, although we are not currently able to attempt a communication with the atmospheric system, it has been possible to rewrite a typical abiogenetic narrative from a different narrative perspective — the author's first-person perspective — in which the emergence of the first living organisms is explained in reference to what they seem to be: a product of mind, the mind of a subject who experiences some form of perception of physical reality and who produces, as typically happens, a significant response to certain interesting stimuli. This narrative, therefore, just like the corresponding classical third-person narrative, has no pretension to be the truth about the emergence of life on Earth; but, while classical abiogenetic stories aim to show a plausible meaningless and purposeless abiogenesis, our rewriting of the story has the opposite goal: to show that a corresponding narrative in purposeful and meaningful terms is possible, although it is not absolutely necessary to explain the emergence of life on Earth, just as it is not necessary to explain the origin of any product of an author's mind in naturalistic terms.

Thus, without any pretension for our narrative to be the true story of the emergence of life on Earth, clearly the previous considerations serve to reinforce doubts around the typical neo-Darwinian interpretation of the phenomenon of life and, consequently, to show that it is legitimate to suspect that today we still do not fully understand what life can be, as suggested by Nagel himself, because we could be completely overlooking a semantic or, more generally, a non-material plane of understanding of the phenomenon; that is, it could admit an understanding of the type we have proposed.

In this regard it is useful to underline that, even though we have identified a plausible natural author and a possible sequence of mental events, this clearly does not prove that life is actually a product of mind, for various reasons. The first of these is that—given our ignorance of the mind-matter problem—just as it is not possible to rigorously exclude that the activity of Earth's atmospheric system is related to subjective experience of mental contents, it is not possible to rigorously state that there is such an experience, and, furthermore, even if such an experience existed, although the activity of the system certainly has a causal power over the emergence of life, it is not at all trivial to prove the intentionality of such an author.

A criticism that could be made of our approach is that it is at least hazardous to hypothesize the appearance of subjective experience in non-biological complex self-organizing systems, because the evolutionary pressure for its emergence would be lacking. But, as we have previously mentioned, one of the factors that make the mind something so mysterious is that, having no causal power over matter, it is extremely difficult to explain in evolutionary terms, because it does not seem to carry any evolutionary advantage, just as it is hard to explain, in evolutionary terms, our interest in mathematics, pure sciences, arts and philosophy. Furthermore, our proposal could be seen as teleological; that is, it could be seen as a particular instance of the natural finalism proposed by Nagel, in that a possible precise aim for the emergence of life on Earth is presented and this is not in accordance with the assumptions of science. The point, in this regard, is that each first-person description is typically teleological, even the first-person description of the origin of a hammer; but although this description is not naturalistic, such an explanation of the appearance remains a naturalistic explanation as long as it is not claimed that the author's mind has a causal power over matter. That is, since at the moment we are unable to explain the experience of our minds, we do not even know how to reconcile the teleological descriptions from a first-person perspective of the creative act that we have described in the previous chapter with a third-person naturalistic description of the chain of physical events (assumed to be purposeless) that occurred in our brain. And, given that it cannot be excluded that the activity of the atmospheric system may correspond to the experience of a mind, as the presence of (living) systems that seem to be the product of a mind suggests, we are effectively equally unable to reconcile the teleological description in the first-person perspective with the naturalistic description in the third-person perspective of the blind process that made life on Earth appear. As we have said, all physical processes are blind by assumption, but there is no way to exclude the possibility that a certain blind physical process underlying the emergence of life on Earth was related to a subjective experience, typically accompanied by a teleological description just like the one we reported in the previous chapter, which is strictly related of course to a blind physical process that took place in our brains.

Thus, the treatment we have proposed on the origin of life is certainly studded with problems and dark aspects, but the problems that it presents are problems that are already present in the dominant neo-Darwinian materialist conception of nature, regardless of whether the experience of mind happens for the first time in biological or non-biological complex systems. That is, the uncertainties and doubts we have about discussing the experience of mind possibly related to the activity of the atmospheric system and, moreover, about considering such a mind capable of elaborating refined ideas such as the contemplation of the beauty of the universe and the intention to represent it are of the same type as those we have with respect to a human brain observed in third-person perspective: why does such an intimate and impalpable thing (as our private subjective experience is) happen to be related to the activity of that complexly arranged physical stuff which constitutes our brains? And how exactly does the propensity to pure and contemplative thinking fit within a conception of life purely in terms of evolutionary advantage?

In this regard, although we cannot prove that life is a product of mind, it must be said that such an approach certainly has one epistemological advantage compared to the typical abiogenetic neo-Darwinian conception: it explains the appearance of design—that patent, mind-resonating quality that life has. And therefore, strictly speaking, even if we cannot prove that life is a product of mind, since life ostensibly seems to be the product of a mind, the burden of proof belongs to those who claim that in fact it is not. As we have seen, a plausible abiogenetic narrative does not constitute evidence at all. Indeed, it has been the starting point for supporting our thesis.

As a final consideration about the whole issue, we just wish to underline that, in our proposal, the emergence of life on Earth is linked to atmospheric little bang events and so, ultimately, to cosmology. As is well known, one of the major current problems in cosmology is constituted by the "Why now?" coincidence, namely, the fact that the present values of the densities of dark energy and dark matter are of the same order of magnitude. Of course, coincidences are not necessarily meaningful, but we just wish to highlight the fact that the dating of the origin of life on Earth is consistent with the dating of the beginning of the energy-dominated era; i.e., while the exponential growth of living systems on Earth was beginning to take place, the universe was beginning its (probably second) exponential expansion. Hence, a "Why 4 billion years ago?" coincidence problem can, in principle, be formulated.

Anyway, regardless of any coincidences, through this approach we have basically attempted to restrict Wilczek's beautiful question (Wilczek 2016) "Does it make sense to consider the physical world as a work of art?" to the biological world, which constitutes the portion that perhaps more evidently seems to be a work of art, and we speculated about who the author might be and what they could have wished to represent. Clearly, in this analysis, an old idea from the history of human thought seems to reappear; i.e. the biological microcosm is seen as a representation of the macrocosm. On the other hand, the idea we have followed may, in a certain sense, also recall the old idea of panspermia, where in this case the seeds of life — mesons with certain characteristics that can also appear in little bang events — spread indiscriminately on all celestial bodies, since they all are targets of cosmic rays. But if a particular target is sufficiently complex to be able to also perform data analysis for event reconstruction in third-person perspective, as well as to mysteriously experience perceptions of such physical events and elaborate on them, we have seen how a reworking of these stimuli leads to a form of life based on chemistry and thus, ultimately, on the electromagnetic interaction rather than on the strong interaction.

To summarize, we have argued—similarly to Nagel—that the mind-matter problem is not a problem limited to humans and probably to some other animal, but has the potential to let us doubt the certainty of the neo-Darwinian conception of the biological world. We have based our arguments on the following reasoning. Since we do not understand how brain activity corresponds to the experience of mind, wherever the dividing line between the so-called 'conscious agents' and 'non-conscious agents' is drawn, it is in fact drawn absolutely arbitrarily, and therefore nobody can rigorously argue how it should be drawn. Therefore, we have illustrated a story in which the line is drawn in an era preceding the emergence of life on Earth, without any pretension of having traced it correctly, but only to show that, if it is traced in this way, motivated by the fact that even the simplest of known living organisms gives the strong impression of being a product of mind, the typical neo-Darwinian conclusions on the origin of life end up being false.

In this regard, we only want to mention that, in reality, the question concerns not only the origins of life but also its entire evolutionary history. It is worth returning to an element in the reasoning process which we have taken for granted but which led us to investigate the possibility that there is subjective experience in natural systems, i.e. systems that are not products of mind. What prompted us to explore this region of systems is that the brain is not considered to be a product of mind; i.e. there is nobody who intentionally modified a pre-existing living organism to obtain the first animal that fully shows cephalization with the typical bilateral symmetry, just as there is nobody who intentionally modified an ancestor of the human to obtain a human, since the process of random mutation and natural selection is a 'blind and unconscious' process, to use Dawkins' words. We just wish to point out that, because of the following points, we cannot be sure of such a conclusion:

1. the atmospheric system is likely to have been active throughout the evolutive story of life on Earth;
2. certain activity of the atmospheric system is a source of ionizing radiation, and the mutagenic power of ionizing radiation has been established;
3. we can't exclude the possibility that there may be subjective experience related to the activity of the atmospheric system.

In other words, we can reconsider any comparative study between the brain and the atmospheric system by arguing, in contrast, that a brain is actually incredibly similar in many aspects to the atmospheric system and, therefore, it is legitimate to question whether biological thinking machines, such as human organisms, are in fact an already existing form of artificial intelligence, produced more or less in the same way as we would somehow produce an artificial thinking machine. In other words, if an author of life exists, their intention could have shifted across time from the one we presented for the origin of life towards the intention of producing a representation of themselves, just as the topics that a human author is interested in often change throughout the course of their life. This too may seem terribly finalistic, but this is not the case. For example, a scientist going to buy electronic components to make a thinking machine, as well as an artist going to buy coloured paints to make their self-portrait, are descriptions of a process in finalistic terms that takes into account the first-person perspective, while from the third-person perspective the purchase of transistors, paints and whatever else is needed is due solely to causes that lie in the past, that is, to previous neural events or, generally, to previous physical events. Therefore, the finalistic character depends on the descriptive point of view. In physics we only know how to study the evolution of a system given the initial conditions, while we have no idea how to rigorously set up a physical problem in finalistic terms; but the fact remains that typically our most significant mental contents are mysteriously, privately experienced in finalistic terms. We don't make any claims about which of the two is the true description, but we can't avoid noticing that the mystery of reconciling these two opposite descriptions is already present in our everyday human life and we are simply underlining the fact that—since humans are part of nature—we can't overlook that this kind of mystery could be far more common in nature than currently considered.

Given the current state of ignorance about the nature of the relation between mind and matter, our discussion thus far has hopefully persuaded the reader to at least consider the possibility that a living primordial organism could be a product of mind, thus casting doubt on neo-Darwinian convictions about the biological world. We now wish to take the thought further to show that the mind-matter problem has an impact not only on the neo-Darwinian conception of the biological world, but also on the whole materialist—or physicalist—conception of nature. We will therefore continue—in the following and last chapter—answering the question underlying this whole work: "How serious is the mind-matter problem?"

Chapter 7

The Hard Problem of Matter

In Chapter 2 we supported Searle's critique of every computational approach to the mind-matter problem, concluding that subjective experience of mental contents should find, in a materialist paradigm, a rigorous explanation in the objective terms of material entities, an explanation therefore expressed in terms of physico-chemical-biological science. An explanation in these terms would be, unlike a computational theory of mind, a rigorous explanation, since it would not presuppose what a theory of this kind must explain: subjective experience itself. In this chapter, therefore, we try to follow a rigorous materialist approach to the mind-matter problem, discussing whether it is possible to find—in principle—a solution to the problem within this approach and, possibly, what kind of solution we can expect to find.

Specifically, we begin by examining, as we promised, Searle's constructive proposal, which we have provisionally taken for granted, i.e. the proposal of explaining subjective experience in terms of causation. In such an explanation, subjective experience of mental contents is claimed to be caused by certain neural activity occurring in the brain. In other words, the question we have left unanswered, and to which we now return, is the following: is it possible to rigorously affirm the existence not only of a generic correlation between the experience of specific mental contents and specific neural patterns, but also of a precise cause-effect relationship between neural patterns and mental contents, where the former causes the latter?

At first glance, the answer is affirmative: many elements seem to suggest that the relationship between brain and mind is a causal relation. In this regard, it is not necessary to invoke the dramatic and famous case of brain damage involving the patient Phineas Gage, followed by many other significant cases of neurological patients in the literature; it is rather sufficient to observe what happens to each of us when we induce a 'slight and temporary damage' in our brain by consuming a glass of wine to suspect that some physical process taking place in the brain is the cause of the corresponding mental states. However, even in these more or less serious cases of brain damage, the only thing there is evidence of is the correlation between neural activity and subjective experience, while the description of the process in terms of causation constitutes a rather spontaneous interpretation of the facts. Such a conception, however, needs to be further motivated, given that up to now we have not at all addressed the question of the nature of mental contents, which is in fact crucial for a description of them in terms of causation. As we will clarify shortly, the thought process necessary to tackle the issue of the brain-mind relationship is different depending on what the nature of the mental contents is assumed to be and, consequently, depending on the reference ontology involved. Such a reference ontology has substantially changed through time; therefore, before addressing the problem in the terms proposed by Searle, it is useful, for completeness, to look at the mind-matter problem in its historical context.

The mind-matter problem has not been presented in these terms since ancient times. On the contrary, it can be found in classical Greek philosophy only from the perspective of the relationship between soul and body. The concept of soul is in fact the concept that comes closest to the current conception of mind, even if the concept of soul was Broader. It did not just include the principle of knowledge; indeed it was also considered the principle of life and movement. As Schrödinger (Schrödinger 1951) observes, the modern concept of force in physics carries within it a residue of animism, while there is no longer any reason to associate the principle of life with vitalism, due to the success of physico-chemical-biological reductionism. Regardless, in Greek antiquity, the main schools of thought on the relationship between soul and body were outlined, establishing the foundations for the ensuing debate around the relationship between mind and matter: the Platonic dualistic solution; the materialist atomist solution of Leucippus and Democritus; and Plotinus's idealist solution. It should be noted that the same speculative directions emerged in Asia, particularly in India: the dualist schools, such as the Dvaita Vedanta and Jainism; the materialistic schools; and the monist schools that we could define as idealist, for example Advaita Vedanta and Buddhism.

Let us return to the Western component of the history of thought, which is more familiar to us. In the wake of Platonic dualism, Descartes affirmed the existence of two ontological substances: the thinking substance (*res cogitans*) and the extended substance (*res extensa*). Descartes, however, identified the thinking substance (spirit, or I, or soul) exclusively with thought and no longer with the principle of life and movement, although for him the meaning of the term 'thought' did not just apply to the pure succession of thoughts, but also included will, imagination and perception, thus paving the way for the modern conception of mind. Beside the work of Descartes, the mind-matter problem owes its modern set-up also to the Galilean distinction between primary qualities (properties that pertain solely to objects, such as mass, extension and so on, independent of any cognitive act of a subject, and that can be studied through the scientific method) and secondary qualities (properties that do not independently pertain to objects, but that derive from a subject's act of cognition, such as colours, smells and so on). The distinction between primary and secondary qualities, combined with the Cartesian identification of mind with thought (therefore with something immaterial), can lead to the conclusion that the mind, having no physical properties because it is immaterial, cannot be studied through the scientific method. Several contemporary anti-reductionist authors will argue that, since the mind cannot be studied through the scientific method because it has secondary properties, the mind cannot be reduced in terms of the natural sciences, or, more radically, that the mind is not a material entity. Regardless, Descartes conceived the mind Platonically as a 'thing', an ontological substance, just as matter is, albeit with distinct characteristics. Let us note, in this regard, that when we introduced the mind-matter problem in Chapter 2, we defined the mind as the set of contents of our experience, but it must be understood that for Descartes the mind is precisely the substance of which thoughts, perceptions and so on are made, just as a ring is made of gold; we can change its shape by melting it and make it become an earring, but the gold the earring is made out of is the same gold the ring was made out of. However, since, for example, your arm effectively rises each time you wish to raise it (unless of course some neural connection is compromised) and, conversely, physical events obviously have effects on mental events, within Cartesian dualism mind and matter, although distinct substances, are united in the human being and interact. And of course such an interaction is the most problematic point of Cartesian dualism: it is not indeed clear how totally different substances can interact with each other. Malebranche's occasionalism—in which God ensures that every time I want to raise my arm it effectively rises—and Leibniz's more refined "pre-established harmony"—whereby the mental and material causal processes run parallel without interacting, but they have been synchronized at the beginning of time by God—are attempts to preserve dualism by avoiding the problem of the interaction between the two substances. In the 1900s, the proposal of epiphenomenalism can be considered a further dualistic attempt to preserve the ontological reality of the mind. Epiphenomenalism holds that both matter and mind are ontological realities, and yet the mind has no causal power over matter, while matter (the brain) has causal power over the mind.

Despite numerous attempts to deal in some way with the problem of the interaction between the two substances, dualism still remained problematic or not entirely satisfactory for several thinkers, and several attempts followed to try to overcome it. Monist approaches have tried to eliminate the problem of the interaction between two different substances by eliminating one of the two substances. To cite some examples, Hobbes proposed discarding the mental substance while retaining the material substance: he proposed a materialist monism in which matter can think. Conversely, Berkeley, and the German idealists later, claimed the existence of the thinking substance only. Idealism starts from the fundamental consideration that whatever we know or will ever be able to know, we know only within our mind; it is not possible to know anything outside our mind. Therefore, even if something existed beyond our mind, it would not be knowable. Idealism states that there is nothing but the mind, i.e. that the only existing reality is the thinking substance, to quote Descartes. Idealism is of course not solipsism: it does not state that what I am perceiving at this moment exists only in my mind. Indeed, when I perceive a table, it does not just disappear once I no longer perceive it. According to idealism, this is because it is perceived within some other mind: the mind of God, for Berkeley; or the superindividual mind, which German idealists call *Geist*, "spirit". With regard to the history of monisms, Spinoza elaborated what will later be defined as neutral monism, that is, a monist ontology in which mind and matter are two aspects—and no longer metaphysical substances—of a single ontological substance, which in Spinoza coincides with God. Spinoza's neutral monism became the 'metaphysical landing' of numerous contemporary anti-reductionists, such as Russell, Nagel and Putnam.

Leaving these less popular monisms in the background, basically the modern approach to the problem appears as an alternative to the approach which attempted to eliminate the problem of the interaction between the two substances of Cartesian dualism. Specifically, it has been trying to remove 'substantiality' from mind; that is, it has been trying to conceive the mind no longer as an ontological substance, but rather as an activity, or a function. This current of thought, whose origins can be traced back to Aristotle, continued with Locke's general critique of the concept of substance, claiming that substances exist insofar as they are what holds together the qualities and what the primary and secondary qualities rest on, but are nevertheless unknowable by the human being. Furthermore, Locke specifically criticized the Cartesian substantiation of the I, followed by Hume, who doubted the very existence of any substance and therefore also of the I as a substance: according to Hume, the I is nothing but a bundle of perceptions, since it is not possible to have perception of one's self independently of a mental content.[1]

However, having discarded the I as substance, the problem of what causes perceptions to have unity arises: they are always experienced by the same subject. Kant will answer this question and will conclude the conceptual critique of the Cartesian substantiation of the I with the introduction of the 'I think', conceived as a unifying function of experiences and no longer as a 'thing'. The identification of the I with a substance, i.e. as a fundamental ontological reality, according to Kant is a conclusion which originates from a paralogism, that is, from a wrong syllogism, which is the following. As stated by Aristotle, everything that is a subject (grammatically, i.e. what is not a predicate) is a substance; the 'I think' is a subject (gnoseologically, i.e. it unifies the perceptions as a subject and is not an object of unification); therefore, the 'I think' is a substance. In the first premise, the subject is conceived in a logical-grammatical sense, while in the second premise it is conceived in a gnoseological sense; the problem lies, according to Kant, in attributing a property of the logical subject to the gnoseological subject. Therefore, he concluded that the 'I think' is not a substance: the 'I think' is a function. This is the contemporary direction of reflections on the mind. The mind is not a substance, but rather a function, an activity, in particular an activity of the brain; the mind becomes a certain activity of a physical system, while matter is the only existing ontological substance.

The anti-metaphysical influence of positivism and later of logical empiricism, jointly with scientific advances in the study of the brain, favoured the emergence of a materialistic monistic perspective. Contemporary philosophical and scientific research on the mind is almost entirely part of a materialistic approach: matter is considered the only ontological substance which everything can be traced back to, while the mind itself is basically identified with the activity of the brain, and it is claimed that matter can think.

However, the materialist theories of mind of the twentieth century were basically born as behaviourist theories. Behaviourism, which emerged between the 1920s and 1950s, emphasizes the observable behaviour of an organism rather than its mental contents, because the latter cannot be studied through the quantitative experimental method. In its most radical versions, it claims that the mind does not exist at all and, therefore, that the mind-body problem is actually a pseudo-problem originating from categorical or linguistic errors. According to extreme behaviourism, the relationship between mind and body is not actually a problem, since the mind does not exist at all: only behaviour exists. These types of behaviourism have been widely criticized, and one of the best-known criticisms refers to the fact that everyone experiences having mental contents distinct from behaviours; basically the point is that I know that I'm thinking, even if there is no relevant behaviour associated with that.

After behaviourism, a new materialist theory appeared in the debate on the mind-matter problem: the theory of identity. It states that the mind—understood as the set of mental contents—is distinct from behaviour and yet its relationship with the activity of the brain is not a problem, because the mind is nothing but the activity of this physical system. Note that this identification is of a theoretical type, of the type "mass is an energy form"; i.e. the identification is accomplished within a theory and is supposedly legitimized by the fact that, through it, it is possible to provide an explanation of the occurrence of certain phenomena. The brain is certainly a physical system, and different activity patterns of the system constitute, in rigorously physical terms, distinct physical states of such a physical system. Such physical states are identified with mental states, which basically are thought of as the 'frames' of our 'interior mental movie'. In other words, when someone experiences the perception of a round shape, such a perception, which constitutes their mental state, is nothing more than the physical state of their brain. In identity theory, therefore, identity is not established between substances, but rather between states of a material system. Therefore, the mind as conceived by Descartes does not exist at all and, consequently, is no longer mentioned in the theory; instead, the term 'mental states' is preferred, and these are claimed to be identical to the physical states of a certain physical system.

Certainly in the materialist theory of identity, the problem of the interaction between the two substances, which afflicts dualism, is avoided by assuming the uniqueness of the ontological substance. However, it is by no means trivial to strictly affirm that mental states are physical states. Typically, in fact, mental states are described through the so-called secondary qualities: colours, tastes, smells and so on; qualities that do not seem to find a place in scientific descriptions. Let us take for example a mental content such as the vision of the colour red. In the physical description of the perception of the colour red what is missing is precisely the colour red: the description in fact concerns an electromagnetic radiation with a wavelength between 500 and 600 nanometres which induces retinal neuron discharge, which in turn induces other discharges in a specific area of the brain, leaving the colour red completely out of the description of the perception process. In the first instance, the mental contents do not seem to be material entities because they cannot be described and studied in the terms and with the methods of the natural sciences. Mental states (characterized by, for example, privacy, teleology, quality) seem indeed to possess different characteristics from physical states (characterized by, for example, localization, causality, measurability), and therefore, with reference to Leibniz's law, it seems immediately possible to conclude that mental states and corresponding brain states are not the same thing. However, several advocates of identity theory have reiterated that the diversity of the descriptions of mental states and of the corresponding brain states could be purely linguistic and, therefore, that it does not imply an ontological and/or empirical distinction. A famous argument in this regard is that even the stars Phosphorus and Hesperus seemed like two distinct stars, whereas today we know that it is just one planet, Venus, appearing as a morning star (Phosphorus) and as an evening star (Hesperus). It is therefore not possible to rigorously affirm nor to strictly exclude the possibility that mental states and brain states are the same entity; however, simply assuming that they are is not sufficient for a full understanding of what identity between the two entities means. As Nagel notes, something is missing from this theory. In his most famous paper (Nagel 1974), he argues that what the theory of identity lacks is precisely the reason for this identification. For this identification to be made with due reason, it must be inserted into a theory that explains how and why mental states are brain states. In the absence of such a theory, their identification remains mysterious to us. As an example, consider that Newtonian physics presents a mysterious equality of the values of the inertial mass and the gravitational mass of bodies; let us call this mysterious equality 'the $m_i - m_g$ problem' and suppose that the problem is raised by a student of physics who is learning Newtonian gravity. If we simply replied that gravitational mass is nothing more than the inertial mass, without making any reference to elevators in motion and so on, regardless of how correct affirming such an identity is, the student would be rather baffled, given that the two masses are conceptually very different: they would not have a theory in which they could insert, understand or criticize such an identification. Moreover, even though the theory of identity seems to be the one that most rigorously preserves materialism, this too is not trivial. If mental states are nothing more than certain particular physical states, the opposite must also be true, i.e. certain particular physical states are nothing more than mental states (if A = B therefore B = A) of a system, which is however physical; an observation that an orthodox physicist would at least struggle to understand.

The problems relative to identity theory have paved the way for the emergence of the computational approaches to the mind-matter problem, which we discussed in Chapter 2. Such approaches arise from a specific critique of the theory of identity: given that we attribute identical mental states (e.g. a pain sensation) to different organisms (e.g. a human and a dog), how is it possible that these mental states are identical to brain states if evidently the brain states of a human being differ from those of a dog? Computational functionalism, which posed this question, does not identify mental states with brain states, but with functional states, distinguishing them from both behaviours and brain states. A mental state is such for the function it performs within the subject's cognitive structure. A mental state is therefore not identified with something material but rather with a logical abstraction, and the irrelevance of the physical realization immediately follows. Criticism of identity theory in this case is called 'the argument for multiple realizability', a term borrowed from computer science, in which the same computer program can run on multiple completely different physical machines. Therefore, in functionalism, and more generally in computational approaches to mind, the relation between mind and brain is often thought of in analogy with the relation between software and hardware.

Perhaps the hardest criticisms of functionalism are referred to the fact that it does not treat mental states satisfactorily, sometimes mistakenly introducing them even when they are not there, and the fact that it does not distinguish between the syntactic and semantic levels, considering the manipulation of symbols sufficient for the appearance of a semantic plane. In this regard we preferred to underline, in Chapter 2, that symbols presuppose subjective experience—since they are symbols for someone—and therefore no explanation of subjective experience in terms of symbols and their manipulation will be satisfactory.

The critiques of computational approaches to the mind led to the appearance, around the 1980s, of a new approach to the problem: connectionism, which can be thought of as a hybrid approach to the problem. Connectionism certainly belongs to the functionalist landscape—thus within it the properties to be taken into consideration in a material system are the organizational ones and not the physical ones—but within connectionism it has been attempted to elaborate models to explain cognitive activities that take into account the functioning of the brain, such as neural network models.

The more recent perspectives on the problem, following on from connectionism, appear to be more refined versions of the theory of identity in which the brain is thought of as part of a larger system—which is basically the whole organism—and are referred to as theories of the embodied mind. In this approach, the types and characteristics of mental states are essentially dependent on physical characteristics of the whole body of the subject who experiences something, in the sense that they play an essential role in causing and constituting the cognitive processes themselves. All mental states share the fact that they are experienced, but in general the specific contents of the experience are different; it is plausible that our perceptions are different from those of a bat, which involve the entire body of the bat and not just its brain. The theory of identity is therefore still among the most accredited theories, even if today identification is no longer restricted to mental states and brain states but is extended to mental states and states of the body, i.e. states of the whole organism.

Having made this brief overview of the problem, we can therefore ask the question we have left open: is it possible to affirm that brain activity causes the experience of mental contents? It is clear that if mental contents are considered for what they appear to be in the first instance, that is, non-material entities, we run into the problem of the interaction between two different substances, typical of Cartesian dualism. Moreover, physics as we know it today seems to constitute a closed causal system in which there seems to be no place for the possible physical effects of a free-acting will, which would violate conservation and/or statistical laws, while the brain seems to respect these laws, like any other physical system, as it is expected to do.

The question seems to be more interesting in the context of the materialistic theory of identity, of which Searle is an advocate: mental states are caused by brain states and mental states are identical to brain states. The two propositions, however, seem irreconcilable, since it is not clear how it is possible to affirm both the identity and causality between A and B. It is in fact possible to say that A causes B if A is different from B. However, Searle does not state that A causes B and that A = B, but that A (brain states) cause B (mental states) and that A and B are both brain states but of different levels: A is a brain state in a low-level description of the brain, while B is a brain state in a high-level description of the brain. Summing up, Searle's proposal is the following: lower-level brain states cause brain states of a higher level which are identical to mental states. Searle therefore states that causality would be between microstates and macrostates of the brain, that is, that it would be a vertical causation, in the sense that low-level states cause high-level states. According to this interpretation, for example, water liquidity is a macro-level emergent event that is causally produced (vertically) from the interaction between water molecules (micro level), and it causally interacts with other macro levels according to a horizontal causation, in which, for example, the condensation of water vapour into liquid water causes the latter to precipitate as rain. However, this type of causation does not convince orthodox physicalists. For example, Jaegwon Kim (Kim 1995) is rightly sceptical about the notion of causation that operates in this account: Searle's horizontal and vertical causation sponsors a confused kind of overdetermination, and his vertical causation does not allow for the time gaps that are required for causal mechanisms to operate.

Although we fully agree with Searle's critique of the computational approach to the mind-matter problem, we do not believe, for the same reasons set out by the aforementioned critics, that the description of the appearance of mental states in terms of causality proposed by Searle is satisfactory, and therefore we cannot find how brain states can cause mental states, nor how and why mental states would be identical to certain brain states. The question is, in fact, of a certain importance because a stance on the matter allows us to discern the type of solution we expect to find to the problem, assuming that there is a solution to the problem in the terms in which it is posed. Searle, indeed, like most of the supporters of identity theory, is confident that the mind-matter problem will be solved in neurobiological terms, that is, that the experience of mental contents will one day be recognized and explained as a biological phenomenon, or as a phenomenon due to the activity of a system with an equivalent causal power and, therefore, explained in terms of the physics of that system.

Whether or not it is justified to believe that mental states are identifiable with brain states and that they are specifically caused by certain brain states themselves, what we want to reflect on is how scientific progress can lead us to the well-motivated identification of mental states with certain physical states in the context of some theory of identity.

Let us therefore suppose that we are in a future in which identity theory is the most accredited theory and that it is actually used to interpret the relationship between mental states and brain states. Let us assume that in such a future the knowledge of the functioning of the brain as a physical machine is complete and that comparative studies between the brain and the atmospheric system have proceeded to such an extent that more and more scholars suspect the possibility that there may be or may have been a subjective experience of mental states related to its activity. The possibility of finding another natural machine whose activity would correspond to subjective experience would be considered, albeit remote, reasonably very attractive with respect to the mind-matter problem. This would in fact allow scientists to shift the focus exclusively from the brain, in which, after all, there is nothing magical or special, even though it might remind one of Aladdin's lamp from which the genie emerges. Rather, some of the brain's characteristics considered crucial for subjective experience to occur may in principle be present in other physical systems, and in this case the atmospheric system could be an attractive candidate. Given the interest in this system, suppose that even the desperate undertaking of communicating with the atmospheric system were attempted, of course with a more advanced technology than our current one. When communication finally occurs successfully, a conversation such as the following might ensue:

Human: "What is it like to be you?"
Planet: "More or less what it is like to be you or anyone else, I suppose, even though the perceptions that each of us experiences might be quite different."
Human: "Just out of curiosity: did you intentionally produce life about 4 billion years ago?"
Planet: "Sorry, it was too long ago, I can't remember."

Let us also suppose that this marvellous natural machine passes some ad hoc version of the Turing test, in which an oblivious subject considers it a human, or an alien, or God, or some other subject of experience. At this point, the scholars involved in the project, having concluded that the atmospheric machine should be considered a thinking machine in the same way as the brain, conduct a more in-depth comparative study between the atmospheric machine and the cerebral machine, which ends with the following assertion:

A physical system that exhibits the following physical characteristics possesses certain physical states which are also mental states:

- it is a self-organizing, complex, non-isolated thermodynamic system that works at a certain distance from the thermodynamic equilibrium;
- its main constituent is water at the liquid state;

- it presents at the lowest level of organization a cellular structure where the cells have dimensions of tens of micrometres;
- it presents at the highest level of organization a division into two parts which present a certain degree of bilateral symmetry;
- the free energy absorbed from its environment, in addition to contributing to the self-structuring of the system, is employed to establish local electric potential across cell surfaces, which under certain conditions discharges, giving rise to pulse trains, which in turn present a certain degree of interdependence, so that the machine presents—as a computing machine—a neural network architecture;
- certain environmental events have a causal power over its discharge activity, and its discharge activity has a causal power over certain environmental events;
- the overall electrical activity of the system gives rise to a global e.m. spectrum that presents the following specific characteristics in terms of its power spectral density...

We certainly cannot know today all the characteristics that the two machines would be discovered to share, but we know that the aforementioned ones would certainly appear in such a study, in which the word "certain" will possibly be replaced by detailed specifications. It should be noted, first of all, that no necessary conditions for the identification of physical states with mental states may appear in the study, but on the other hand the individuation of a sufficient condition—which occurs when the system possesses each of the specified characteristics—presents several problems as well. Basically, even admitting that mental states coincide with some physical states of certain physical systems—certainly with some brain states—it is clear that the theory does not explain at all why only some physical states of some physical systems made in a certain way are mental states.

This is a considerable problem, which also appears in terms of causation, and is linked to Chalmers' critique about the possibility of discovering any physical process identifiable as the cause of subjective experience: why is a certain specific physical process also a certain mental process? Why should a physical process give rise to subjective experience? Objectively, it seems unmotivated that this should happen and yet it seems to happen. It is therefore not enough for us to know which process gives rise to experience: we need a description of why and how; otherwise, without an explanation of why and how a specific physical process gives rise to subjective experience, we are condemned to affirm that it is so because it is so. In both systems considered in the previous thought-experiment, certainly the physical field that plays a crucial role is the electromagnetic field, since the physical states of both systems are determined by the discharge patterns of the electric potential. In practice it is not clear what it means that a certain dynamic configuration of the e.m. field, which is represented in principle by a solution of Maxwell's equations, is a mental state, as well as being the physical state of a certain physical object, i.e. of the electromagnetic field.

The dissatisfaction with the answers provided by the theory of identity certainly does not end here. In the context of the mind-matter problem, it is not only a question of explaining the nature of the experience of mental states in physical terms, but also of explaining the specific relationship between a brain state and a mental state. To explain what we mean, we assume that we are not subjected to some version of Donovan's experiment; i.e. we assume that when we experience, for example, the perception of a round shape, there is actually a round shape in the physical world. Let us say for simplicity that every time a round visual stimulus is present in front of our eyes, a square discharge pattern forms in the visual cortex. In correlation with the formation of this square pattern, we experience the perception of the round shape which is physically in front of our eyes. Let us note that the perception of the circle does not depend on the fact that there effectively is a circle in the physical world in front of our eyes, but on the fact that the square pattern is formed in our brain. It is also reasonable, by virtue of various transcranial stimulation experiments—more feasible compared to Donovan's famous experiment—that if that square pattern were induced in some way other than through natural retinal stimulation, the subject would invariably experience a circle. The problem, posed in a very specific way, is: why do we perceive a circle—which we believe to actually exist in the physical world—precisely when that specific square pattern is formed in the visual cortex? What is the relationship between that specific brain pattern and that specific perception? Why don't we see a circle, for example, when the neural patterns are circular or triangular? As we mentioned in Chapter 2, it is possible to identify rules of correspondence between neural patterns and perceptions, as if perceptions were somehow encoded according to specific neural patterns. The problem is that, as much as we try to support the theory of identity, we do not understand at all the nature of the coding mechanism within neurophysiology. Conversely, we have a full understanding within biology of what it means that proteins are encoded in DNA, since in the latter case the coding rules are established by the specific aminoacyl-tRNA synthetase proteins, which are encoded in the DNA itself like all the other proteins. Therefore, it is absolutely clear what we refer to in biochemical, and therefore physical terms, when we use the term 'genetic code'; while the theory of identity, like the other materialist theories of mind, does not provide us with any clues on the nature of this mysterious code of perceptions. Thus, according to the proposed thought-experiment, although we are confident in the progress of the sciences, we believe that what we can expect is that the physical processes possibly related to subjective experiences will be discovered and analysed in greater and greater detail, but that the discovery of these details cannot be significant for solving the mind-matter problem.

Having summarized the history of the approaches to the mind-matter problem, paying particular attention to the materialist approach that has led research in this area at least in the last century, because of the difficulty in finding a rigorous and definitive solution within this approach, we believe it is useful to investigate the concept of matter in more detail: what exactly do we mean when we refer to matter or to material/ physical entities? Up to now we have in fact taken for granted that we know what matter is. In this regard, one of the most accepted definitions of matter is the following: matter is "what the best physical theories claim it is", but even this definition is not without its problems, because it has to deal with Hempel's dilemma. Hempel argued that physicalism is either false or incomplete or indeterminate: the best current physical theories are either false—because they will be subjected to even more radical revision in the future—or incomplete, while true and complete future physical theories are obviously indeterminate in content. Therefore, the best definitions of matter that the physical sciences can offer could be either false, or incomplete, or indeterminate in content. Giving a good definition of materialism and material entity is therefore also an arduous task, and the lack of a good definition of one of the two terms has a significant effect on the mind-matter problem. We believe that one of the factors that contribute to making the contemporary approach to the mind-matter problem so intricate is that, while it is very clear what we mean by experience of mental contents, it is not so clear how matter can be defined. This is not surprising given the epistemological primacy of the mind. When we talk about mental contents we know by direct experience what we are talking about, while when we talk about matter we are talking about something that has never been a mental content in itself except as a concept. At this point, it is useful to present a plausible reconstruction of the thought process that leads to the birth of the concept of matter, both in psychological and philosophical terms.

Let us start with the epistemological primacy of subjective experience. Since mental contents are defined as the only 'things' of which we have direct experience, they are the only 'things' whose existence we are certain of. This means that no one can rigorously affirm — or deny — the existence of anything outside their own mind, and neither can they be certain of its nature. Despite this leaving room for radical scepticism, non-agnostic positions have arisen. The solipsist position denies the existence of something else outside one's own mind, while the realist position, on the contrary, affirms the existence of a world outside your own mind. While the solipsist position is a rare attitude and usually the result of thoughtful reflection, the realist position is adopted not only deliberately, but also spontaneously, unconsciously: it is part of so-called common sense. The realist position is therefore the most commonly adopted, consciously or not. Even if one cannot be sure of the existence of something else outside one's own mind, one typically lives with the — naive, in most cases — belief that it is so. On the other hand, even if radical scepticism is absolutely legitimate, in an attempt to avoid it still today, naive realism has not been completely abandoned. For example, both Putnam and Searle are supporters of naive realism, in which it is claimed that the contents of our perceptual experience are nothing but objects of the real world. Naive realism constitutes a position which, although widespread, is so difficult to uphold rigorously that we avoid taking it into account, consequently accepting the legitimacy of radical scepticism regarding what exists beyond our minds. Therefore, assuming that indirect realism is true, we consider that typically there is no awareness of the fact that we experience mental contents and not the 'real' world outside the mind of every one of us. Certain mental contents are confused with entities of the world, even by those who have thought about it and believe, for example, that the perception of an apple is not that apple; perhaps those who suffer from vision defects such as myopia and astigmatism are facilitated in distinguishing reality from its perception. The mental contents that are confused with entities of the world are characterized by persistence (you perceive the Sun in the sky every single day, without exception), by coherence (they follow regularities), by immutability (you cannot perceive a green Sun, even if you wanted to), and by the fact that you typically perceive them as outside yourself. Other mental contents, such as mental images, thoughts or dreams, may not have the same persistence or coherence, are modifiable by the subject who thinks of them, and are not typically perceived as external by the subject; rather they seem to be perceived internally, inside oneself. By virtue of these differences, therefore, the latter are actually considered mental contents, while the former are— erroneously—not classified as mental contents, but as entities of the world, existing outside the mind. If these are not mental contents but rather entities of the world, they must therefore be made of something that, in a still undetermined sense, is 'not mental', generally referred to by the term 'matter'. The entities of the world are material entities as much as the body of the subject who conceives the concept of matter. Furthermore, those mental contents confused with entities of the world, like the Sun we see in the sky during the day, are considered entities of the world because they are perceived by several subjects. There is therefore an intersubjective coherence between perceptions, reinforcing, among others, the conviction that the Sun is an entity of the real world, because it doesn't seem to be characterized by privateness, but by publicity.

Congruently with such a reconstruction of the origin of the concept of matter, animism was probably the first way of interpreting reality. Not only animals, but also plants, the Earth, the Moon, the planets, the Sun and other stars, were each considered to be animated by a soul. According to Schrödinger (Schrödinger 1951), it is thanks to Greek philosophy that the West has overcome animism: only animals were considered by the Greeks to be animated by a soul, while all other entities of the world were considered to be ruled by natural laws. This conception constitutes the foundations for the formulation of Cartesian dualism, which is, by the way, the dominant conception in modern conventional wisdom. Conventional wisdom trusts that there are two realities: a mental (or spiritual) and a non-mental, material reality. In the human being, these two realities coexist, so that the human body is subjected to both natural laws and a mental will. The interaction between the two realities is considered to be common sense, since it seems obvious that the mind can have power over matter: I decide to raise my arm and it rises. While common sense does not go beyond that, philosophical reflection can attempt to overcome the problems of ontological dualism, as previously mentioned, or it can consider the interaction between two different realities too problematic, gravitating towards monism, by considering one of the two realities as the only existing one.

This plausible reconstruction of the birth of the concept of matter and of the ensuing ontologies aims to show that the concept of matter is presented in a rather spontaneous way according to a *via negationis*: matter is what is non-mental. Clearly, what is meant by "non-mental" needs to be better specified. According to a *via negationis*, matter can be thus so defined: matter is the substance of which anything that exists independently of being known within a subjective experience is made; or, in short, matter is the constitutive substance of whatever exists in purely objective terms. More precisely, in natural science we are typically interested in objectively studying the evolution in time of the states of a physical system; therefore matter must be defined as the stuff of which the things that exist and are in a certain state independently of being known are made. For example, Earth—a material object—exists and occupies a certain orbit around the Sun, a certain position at a given time, regardless of whether or not these are known. Given such a definition of matter, it is easy to see that this concept fits perfectly with the scientific method of investigation of reality. It is therefore no wonder that, in the contemporary age, in which philosophy is largely aimed at supporting the scientific method, materialist monism is the most adopted philosophical position and is also the paradigm within which scientific research is carried out. The materialist position affirms that material entities—those naively confused with mental entities or consciously postulated as existing outside of minds—are the true reality which everything else must be reconducted to; even mind. The prestige of the scientific method, due to its success in understanding nature, has led to forms of naive scientism, in which only the existence of the entities studied by the natural sciences is affirmed. Conversely, because of their private nature, mental contents cannot be studied in the third-person perspective through the scientific method, so the mind has never been observed in the study of nature, since it is unobservable, or undetectable. Therefore, it is not surprising that with certain forms of behaviourism an attempt has been made to deny its existence.

The problem that the materialist position brings with it is quite evident: if there is nothing but matter, how can the nature of mind be material? If it is assumed that matter is the only ontological substance, this must be the substance of which the brain system is made; moreover, certain states of the brain are identified with mental states in the theory of identity. The problem is that mental states do not exist independently of being known, and this contradicts the very definition of matter.

At this point, for materialism to be true, the existence of mental states should have to be denied. In fact, it could be said that eliminative materialism is the only rigorous materialist perspective. According to eliminative materialism, of which some forms of behaviourism were precursors, mental states do not exist, but only brain states exist; according to eliminativism, the nature of the problem is purely linguistic. Even though we use two different types of language—a neurological one to refer to brain states and a psychological one to refer to mental states—this does not at all imply that we are referring to two different ontological entities, but rather that psychological language is an unscientific language which, in the future, may be completely supplanted by neurological language. Once matter is assumed as a fundamental reality, everything must be reduced or somehow traced back to it—unless one wants to abandon materialism itself—and mental states are no exception.

However, subjective experience has a peculiarity, and it is this peculiarity that draws so much interest and that makes it resist the various reductionist materialist attempts—namely, the epistemological primacy. As has been evident, at least from Descartes onwards, the only certainty we possess is the existence of our subjective, or mental, experience. Everything else is derived from this, even the concept of matter. Thus, affirming that matter exists, and that subjective experience does not, is an operation whose legitimacy is hard to sustain, to say the least. It is not at all satisfactory to deny that the problem exists, discarding the existence of subjective experience.

Given the difficulties that the problem poses, even though we have already briefly tried to tackle them from diverse perspectives, it is appropriate to further focus on the root of these difficulties: why is the mind-matter problem so hard? Or, to put it in the terms introduced by Chalmers: why is there a hard problem of consciousness? Up till now we didn't need to refer to the term 'consciousness' to address the problem, and we have preferred to avoid it so far because it can be a misleading concept, still lacking a widely accepted definition. However, since we now intend to refer to positions of other authors who use this term, we cannot refrain from presenting a definition that works for us and which is coherent with the meaning attributed by other authors. We define consciousness as an inherent property of mental contents, just as, for example, privateness is a property of mental contents. Specifically, by the term 'consciousness' we mean a trivial property of mental contents which consists in the fact that they exist insofar as they are experienced. In other words, by definition, mental contents exist if and only if they are experienced, or, alternatively, if and only if they are known. Given this definition, the answer to the question "Why is consciousness so hard to explain within materialism?" is quite obvious, if the definitions we provided both for consciousness and for matter are compared: matter is the stuff of which everything that exists and is in a certain state independently of being known is made, while consciousness is the fundamental property of mental contents, which exist only as long as they are known. And certainly, thinking of mental contents as states of a material system does not provide any advantage, since a material system—by definition, made of matter—is assumed to be in a certain state regardless of whether that state is known by a subject of experience. In this regard, Nagel, in his famous paper (Nagel 1974), affirms the difficulty of the scientific method of studying what is subjective, since while objective knowledge is such because it deviates as much as possible from the specific human point of view, subjective experience is instead strictly pertinent to a specific point of view; eliminating that point of view in order to arrive at an objective knowledge of consciousness is a dead end that does not address the real problem: is it possible to objectively study a subjective point of view? Nagel, in his paper, concludes that physicalism is not necessarily false, but we have no idea how it could be true. Similarly, Hutto (Hutto 1998) affirms that the difficulty of the mind-matter problem lies in the fact that it does not fit into the classic scheme in which physical phenomena are included: it does not fit into an object-based scheme. Hutto points out not only that consciousness does not fit into this scheme, but also that not all physical phenomena are understandable through it: the phenomena studied by quantum mechanics, for example, are not comprehensible through it. As already pointed out by Schrödinger (Schrödinger 1951), by dividing reality into subjects who observe an objective world, and giving science the task of studying the objective world, the subject remains outside the field of science, and it is very difficult, if not impossible, to recover it at a later time. Shrödinger went much further, saying that dual language in terms of subject and object is misleading; in saying this he was probably inspired, like others of his time, by the philosophical tradition of Advaita Vedanta, the Eastern philosophical tradition that comes closest to some formulations of Western idealism. Schrödinger recognizes in the postulate of intelligibility—natural phenomena are understandable through laws of nature—and in that of objectification—the hypothesis of the existence of the real external world and the exclusion of the knowing subject from the desired intelligible image of the world—the postulates assumed by scientific thought since its inception, or rather, thanks to which it was born. He remarks that these postulates are not necessary as such for scientific progress; they were, rather, the products of a first unconscious simplification that allowed incredible scientific successes, but they also carried terrible antinomies in their womb, such as the mind-matter problem.

Since the mind-matter problem presents difficulties that are at least comparable to the difficulties that arise when trying to find a satisfactory interpretation of quantum mechanics within the familiar objective view of the world, there have been a plethora of theories trying to establish a metaphysical link between these two strange types of phenomena. In this regard, Chalmers observes that the attractiveness of quantum theories of consciousness may derive from a "law of minimization of mystery". Consciousness is mysterious, and quantum mechanics is mysterious in a similar way: they both challenge the postulate of objectivity. Then maybe the two mysteries have a common source. We are sceptical about such theories of consciousness, but we do believe, in line with Hutto (Hutto 1998), that there is a connection between these puzzling phenomena which can be deployed to alleviate the two mysteries. What makes both these phenomena peculiar is that neither fits easily within an object-based schema. In other words, neither sits happily in a conceptual schema that derives from our responses to a world of macro-sized, spatio-temporal objects, however extended and developed by theoretical considerations.

Therefore, within materialism, not only the very existence of mind seems to be unintelligible, but, in fact, not even matter itself seems to be intelligible as we have previously conceived it. Our position, therefore, is that, given these problems of unintelligibility, unless we want to accept the condition of unintelligibility every time it arises, it would be appropriate to explore the possibility that the null-hypothesis of matter as we conceived it can be rejected. That is, what Chalmers defines as the "hard problem of consciousness" is ultimately due to the assumption of the exclusive existence of an ontological substance — matter — of which the things that exist in a certain state regardless of being known are made. Since we are more certain of the existence of mental contents — which instead exist only as they are known — we prefer to refer to the hard problem of consciousness as the "hard problem of matter". This is the problem that appears among our mental contents when we assume that the only fundamental ontological substance constitutes objects that exist independently of being known, thus introducing an unbridgeable gap between the ontological and the epistemic plane, i.e. between the plane of being and that of being known. In other words, materialism would be fine if there were no one to know it, and therefore—since we are not philosophical zombies—we have no idea of how materialism could be true.

This work could therefore end here, since we have finished the arguments supporting the thesis that the mind-matter problem is much broader and more serious than it is typically thought to be and has the potential to challenge not only the current conception of the phenomenon of life but also our whole comprehension of nature in terms of the concept of matter. Nonetheless, although the purpose of this paper is not to take a position on the mind-matter problem, in a work of this type it is typically expected that authors do take a position. We therefore dedicate these last pages to illustrating, by way of example, how the problem could be posed from a different perspective, which seems to us more tractable than how it is currently set up. Having sustained the unintelligibility of the relationship between mind and matter in certain ontologies, our starting position is aligned with that proposed by Hutto (Hutto 1998): the problem, since it cannot be solved, must be avoided; that is, every ontology in which the problem appears must be avoided, such as dualism and materialism. In other words, any metaphysics in which matter is considered ontological should be avoided, because it is in this way that the insoluble problem of the relation between mind and matter appears.

In this regard, let us take up Nagel's critique according to which the identity makes sense in the context of a theory. To make a comparison, we can compare the mind-matter problem to the problem that arises within the framework of Newtonian theory, in which the mysterious identity between the values of two distinct physical quantities, such as gravitational and inertial mass, is not explained. And today we know well that there is no way to explain it in that conceptual framework, where, in our analogy, the Newtonian theory of gravity represents materialism as a theory of reality, and the mysterious relationship between gravitational mass and inertial mass (the '$m_i - m_g$ problem') represents the mind-matter problem. No theory of the identity between the two physical quantities that leaves the Newtonian theoretical framework unchanged adds anything to our understanding of the $m_i - m_g$ relation, and the same happens within the theory of the identity between mental states and brain states, assuming that one wants to accept the possibility of such an identity. What would really solve the problem would be some deeper theory of gravity where the problem is avoided. The relationship between the two masses would be explained through a different conceptual framework — some other ontology — in which the previously accepted one takes on the character of appearance, if not illusion, just as has happened with the force of gravity. It is therefore a question of finding some ontology (which in the analogy is general relativity) in which the concept of matter as a non-mental substance (which in the analogy is the gravitational mass as the source of the gravitational force) is apparent, illusory; that is, it is not what it seems to be. We need an ontology therefore for which the mind-matter problem — like the problem of the relation between inertial mass and gravitational mass — does not exist. The ontology we want to arrive at must necessarily be some form of ontology in which matter does not appear as an ontological substance, therefore an ontology that eschews Cartesian dualism, as well as materialism of course. Such an ontology could therefore have a place within some form of neutral monism or idealism, but in fact it is so difficult for us to take on neutral monism — due to the mysteriousness of this substance which manifests itself according to a mental and a material aspect — that we can only discuss the only option left: some form of idealism, in which what is typically conceived as material must ultimately be mental. Some form of idealism would constitute the only ontology that we can conceive in which the fracture between the epistemic and ontological planes, which is ultimately responsible for the mind-matter problem within materialism and dualism, does not occur. Perhaps Planck (alongside many other founders of quantum mechanics) had an insight into the necessity of doubting materialism itself when he said: "I regard consciousness as fundamental. I regard matter as derivative from consciousness."[2] And it is possible that even today the ontological perspective of the founders of quantum mechanics has not yet been fully appreciated. We suggest that the impossibility of getting out of the mind-matter problem prompts us to reconsider that part of their thinking which was accused at the time of being mysticism. At this point, we believe it is clear that for us the mind-matter problem cannot be addressed solely in the context of the scientific method and lends itself to broader philosophical speculation, from which we cannot expect one definitive answer, but just answers that everyone can find more or less reasonable, especially since there is no experiment that can reveal the actual existence of matter.

In the idealist conception, the mind is the only fundamental reality from which everything else derives. It is possible to identify Plotinus as one of the first Western monist idealists. He indicates the fundamental reality with the term "One": the One, through intuiting itself, emanates—by superabundance—existence. The emanation takes place by stages—hypostases—which have an ever lower degree of reality and greater degree of multiplicity. The One, by self-intuiting, creates the Intelligence—the Nous—which contains in itself the first duality: that between thinker and thought, i.e. between subject and object. The Nous, in turn, by thinking, creates the Soul, which is a cosmic, universal Soul that contains the multiplicity of souls. Matter, on the other hand, is thought of by Plotinus as the absence of being; therefore it does not exist in itself, but only as a privation of reality. Conversely, Berkeley's idealism is a pluralistic idealism. Berkeley argues that the only fundamental reality is spirit (mind), and affirms the existence of an infinite spirit (God) and a plurality of finite spirits (human beings); ideas and finite spirits are produced by God. The spiritual substance is active and perceives the ideas that it produces. We—humans—perceive the ideas produced by God and the ideas produced by our minds. Everything we perceive has only these two possible origins: the mind of God or our mind. There is no other kind of substance other than the spiritual one, and both God's ideas and ours exist only as being perceived; therefore an idea exists only within a mind. There is an ontological difference between spirit—or mind—and ideas. Ideas are caused, produced by the spirit, and are not a substance; their being consists in being perceived. Ideas are therefore passive: an idea is not the cause of another idea; only the spirit is the cause of ideas. Within this system of thought, Berkeley explains the intersubjective agreement between perceptions in this way: that which we all perceive, like the Sun in the sky, is not an idea produced by every single mind, but is an idea produced by the mind of God. In this way the idea is unique, and therefore perceived in the same way by everybody.

Similarly, according to Leibniz, monads—which are psychic indivisible units—are created directly by God and are indestructible; monads are complete entities, and each of them reflects the whole creation from a particular point of view, while God collects within all the infinite points of view. Monads are mutually independent, i.e. they are "without doors and without windows"; it is God who, from the beginning of time, has established a coordination among monads, the so-called "pre-established harmony", which, in fact, constitutes Leibniz's solution to the problem of the intersubjective agreement between perceptions. Perceiving something, according to Leibniz, does not, however, mean "being aware of it": conscious perception is instead what Leibniz refers to as "apperception".

Only the aggregates of monads organized by a soul (which is also a monad) are endowed with apperception, that is, with conscious perception. The aggregates of monads not organized by a soul have perceptions of reality without, however, being aware of them. A stone, for example, is an aggregate of monads that are spiritual substances, and therefore it perceives, but it is not conscious because it is not organized by a soul. Thus, all monads and aggregates of monads do perceive, but only a few of them are aware of their perceptions. What is typically referred to as matter, according to Leibniz, are the unconscious monads and the aggregates of unconscious monads, which lack apperception. Leibniz does not want to deny the mechanism of his era, but he wants to suggest that mechanism is true in a narrow, superficial context, since it veils a much deeper and wider spiritual reality.

According to Fichte, instead, the I is the first principle of reality, a self-creating activity (*Tathandlung*), both the actor and the product of its very activity; i.e. the I posits itself. The I not only posits itself, but also opposes itself, in itself, the non-I, a term that for the philosopher is synonymous with *object, nature, matter*. The I, having posited the non-I, comes to be limited by it, and for this reason it finds itself existing in the form of a multiple and finite I; human beings are finite Is. Fichte's philosophy is a sort of ideal-realism: idealism because it claims that the non-I is a product of the I, and realism because it admits an influence of the non-I on the I. According to Schelling, on the other hand, the infinite principle which is the creator of reality cannot be reduced either to the subject (Fichte's I) or to the object (the non-I); rather it is the foundation of both, the unity or identity of both. Finally, Hegel, the most famous of the German idealists, affirms that reality is the Spirit, the Absolute, the Subject, the Reason, the Infinite in which all finite reality is resolved; in practice it affirms the existence of a single supra-individual mind. The Absolute is understood in a pantheistic sense, as an immanent reality in the world, and is infinite as a self-sufficient Totality in which every finite reality is resolved. The finite as such does not exist, as it is a partial expression of the Infinite; it is therefore the Infinite itself. The Infinite is not static, but rather it is an evolving reality that produces itself through a dialectical process, in which the dialectic is at the same time the law of development of reality and the law of understanding it. The Infinite is Reason insofar as, for Hegel, "what is rational is real, what is real is rational", a panlogism in which the full identity between what is real and what is rational is affirmed.

Hegel's idealism is perhaps the best known in the history of Western thought, and clearly the question it immediately raises is the following: how does the hypothesis of a single supra-individual mind reconcile with the apparent multiplicity of minds, i.e. of my mind, your mind, and so on? Let us note that, similarly, in materialism we typically refer to a plurality of bodies in the material universe; certainly my body, your body and, for example, celestial bodies are just part of the material universe. Moreover, as we underlined in Chapter 1, it is certainly more accurate to think of our bodies as thermodynamic systems rather than as bodies; those that are identified as our physical bodies by common sense are in fact much more similar to a burning flame, a hurricane, to a dissipative flow of matter and energy, rather than to a stone. Thus, all material bodies, and even more so our bodies, exist only apparently as individual entities, since they are nothing but part of nature: the particles our bodies are made of are excitations of the same physical fields whose excitations constitute everything else. Would it make sense, then, to similarly introduce a universal subject—let us say God—and to claim that our private minds are just part of the mind of God? Basically, we are referring to what Schrödinger referred to when affirming the unity of consciousness, but we have preferred—for clarity—to explicitly introduce a single universal subject: God. Of course with this term we are certainly not referring to a theistic God, such as the one invoked by creationists in science-faith debates: we have already extensively expressed our critique of neo-Darwinian conclusions not in terms of religious beliefs, but in terms of the mind-matter problem. Similarly, here we are criticizing the concept of matter upon which modern physics is founded, and we are not saying that there is anything wrong with physical theories, other than that they will be replaced by better theories as science develops. What we are discussing here is the interpretation of what physics truly studies and, consequently, the higher-level sciences. Thus, the God we are referring to has nothing to do with science-faith debates about "Darwin versus religion" or "multiverse versus fine-tuned universe", but rather with our conception of what is real, and of course—for those who prefer it—an alternative terminology in which the word "God" is avoided can easily be found.

Anyway, in this type of idealism, matter—the non-mental substance—would be replaced by an ontological constitutive substance, which would be the very substance our mental contents are made of: the stuff of which the so-called physical world is typically assumed to be made would be the same substance of which our thoughts, perceptions and so on are made. With the introduction of such an ontological substance, the mind-matter problem needs to be reformulated in idealistic terms: what would be the relationship between the mental contents of God—the objects of the so-called physical world, including the brain—and the private mental contents of each of us?

Obviously we cannot expect to give a definitive answer, nor can we expect to give an exhaustive treatment, but we just want to show how in idealism the problems that emerge seem to be more treatable than the unintelligible mind-matter problem.

Suppose we build a computing machine that performs data analysis for event reconstruction, equipped with some kind of particle detector. The machine will present a certain number of output channels whose physical parameters— tension, frequency, and so on—encode in a way that is known to us (since we built the machine) the values of some physical parameters of the revealed event, such as the particle momentum, and so on. Let us also suppose that we can wear this machine on our back as a backpack and connect the output channels to the skin on our back, so that the physical output parameters of the machine are perceivable as trains of electrical stimulation. So let us suppose we spend a long enough time training ourselves to use this machine, whose human-machine interface is constituted by such discharge patterns on our skin. Since we have built it, we know how the outputs are encoded, and with the continuous use of the machine we find it less and less difficult to build our mental image of an event starting from the response of the machine to the events. While using the machine, which in fact constitutes an appendage of our body and brain, a sort of prosthesis, we experience two types of mental contents, which in semiotics are called "signified" and "signifier", where the signified pertains to the plane of content, while the signifier pertains to the plane of expression. These are the two main planes of a sign. In linguistics they are the plane of the word as a sequence of letters and the plane of the meaning which the sequence points to. The code is what establishes the relationship between the two planes; e.g. the genetic code establishes the rules of correspondence between certain triples of bases (signifiers) and a specific amino acid (signified). Within a language, syntax describes the rules by which signifiers can be combined into sentences, while semantics describes what is signified, what they mean. In our case the signifiers will be constituted by, for example, the levels of the machine-output voltages that we can experience as perceivable electric discharges on our body, while a signified will be constituted by certain physical characteristics encoded by precise machine-output patterns according to coding rules that we know because we built the machine.

Now suppose that we did not build such a machine and we are not the ones who constantly feed it with some source of free energy, but that the machine appears as a self-organizing system at thermodynamic disequilibrium, like for example the atmospheric system which responds to cosmic ray events, or like each of our brains, which respond to photons impacting the retinal neurons. What happens in idealistic terms in such cases is that a part of all the mental contents of which reality is made—that is, the part that will constitute Earth's atmospheric system or a human brain—is being organized in order to detect and elaborate stimuli that pertain to the remainder, producing output patterns according to a known encoding. The only way we can imagine to explain the relationship between each of these systems and someone's private mental contents is therefore this: the possible mental contents of Earth's private experience are mental contents, just as the whole atmospheric system, the cosmic rays that stimulate it, and everything else that exists are mental contents; but among the whole ensemble of mental contents that the universe is made out of, some of that—lightning discharge patterns—are signifiers, to which a corresponding set of mental contents appear, which are the correspondent signifieds. Such signifieds are the mental contents that constitute what we refer to as the mental contents of someone's private experience: in this case the possible private experience of Earth, thought of as a subject of experience. It is of course possible to express the same interpretation of what the mental contents of each of us are. Nature is constituted by God's mental contents, and your mental contents are just a subset of God's mental contents: they are the signifieds that correspond to signifiers constituted by your brain discharge patterns, which of course are God's mental contents just as everything else is.

We cannot certainly claim to have produced an explanation of the relationship between mental contents and brain states. We have presented the previous naive interpretation of such a relation just to express that we believe that in ideal-realism the mind-matter problem is more tractable than in materialism; that is, if in materialism we are faced with an unintelligible problem, since basically consciousness is irreducible within materialism, in non-pluralistic idealism we are faced with more tractable problems, such as the problem of multiple minds, i.e. given that nothing but a universal mind exists, why does it seem that there are multiple minds?[3] We suggest that indeed this problem could actually be ascribed to the limits of language, while it is untenable that the mind-matter problem within materialism is a pseudo-problem due to language. Specifically, we are suggesting that the mysterious code of perceptions is what is responsible for the Veil of Maya, as formulated by Schopenhauer, that sets an apparent boundary which supposedly distinguishes human minds, and probably animals' minds and Earth's mind, from God's mind. The Veil of Maya, the code that introduces a distinction between what things are (as symbols or signifiers) and what they represent (the signifieds or meanings), is ultimately responsible for the veiling of the nature of the products of mind, in that they are not completely understandable in terms of what they are, but they admit a semantic plane of understanding. Specifically, this is the argument we used to argue that the first living organisms could actually be a product of mind, since it is possible to identify in them a semantic plane of understanding.

Our position with respect to the problem is therefore that a form of ideal-realism such as the one we have mentioned constitutes the most suitable ontology to avoid the mind-matter problem. It is useful to note that a monism of this type is not all that different from a rigorous materialism: the fact that we look at the universe and reflect on it is expressed in materialism by saying that matter is looking and reflecting on itself from a localized perspective. Moreover, the answer that materialism seems to provide regarding the problem of free will also remains fundamentally the same: since there are no actual individual entities, there is no real individual free will. In this regard we just wish to note here that in any non-pluralistic ontology it is difficult to claim the existence of free will, since free will is typically conceived as individual free will, i.e. the free will of someone, but in non-pluralistic ontologies the plurality of individuals does not really exist. We therefore believe that such an approach to the problem of free will can be more fruitful than the typical approach in terms of "determinism versus free will". Conversely, what radically changes between materialism and ideal-realism is the perspective from which we look at the hard problem of consciousness. Within materialism, what is happening when, for example, a physicist formulates a quantum field theory, is that fundamental physical fields—or something similar that we conceptualize in these terms—are reflecting on themselves from a localized perspective, even though how this is possible seems absolutely unintelligible; or, equivalently, consciousness seems irreducible in physical terms, and we struggle with the impossibility of psychophysical reductionism. In this regard we have proposed that, instead of struggling with the impossibility of psychophysical reductionism, the relation of a human—such as an artist or a scientist—with nature is far more easily conceivable in terms of mind that reflects upon itself rather than in terms of matter that reflects upon itself. Basically, in idealism, laws of nature are only recognized in a human mind, but they exist as mental contents in the mind of God, which human minds belong to. Such a conception of reality, in addition to making the mind-matter problem more manageable and quantum mechanics more smoothly interpretable, overcomes another age-old problem of materialism: what are the laws of nature in themselves, besides being mental contents within physicists' minds? The question is certainly connected with the postulate of intelligibility, implicitly mentioned by Einstein in his famous observation: "The most incomprehensible thing about the universe is that it is comprehensible."[4] In ideal-realism there is no need to explicitly assume the postulate of intelligibility to be true, since it is an epistemic consequence of ontology itself, while the postulate of objectivity is discarded, given that there are no objects in such an ontology that are independent of the subjective experience of them.

Certainly, such a discussion cannot be exhausted in the previous few lines, but in conclusion we want to emphasize that, in facing such problems, the individual psychological factor regarding the acceptance of a vision of the world cannot be neglected. As much as it is difficult for most of us to accept the absence of individual free will, it is at least as difficult to accept the possibility that this hard and massive stuff of which the real world is made (which we call matter) does not actually exist as currently conceived.

In the conclusion of a work like this, which revolves entirely around the subject of subjective experience, we find it useful, in an attempt to make the aforementioned psychological difficulties more surmountable, to report a subjective experience: specifically, a dream, a dream that anyone who has dealt with the hard problem of consciousness and with that of free will, with due stubbornness while awake, could probably have:

You are a sort of monkey who lives in the jungle with its group, but, for some reason you can't remember, you have strayed too far from your group, and as the evening falls you have lost your way back to the group you belong to. So, you walk alone in the jungle during that dark night, scared by sudden flashes and by the overall stormy atmosphere. As the end of the night approaches, the fair weather begins and your fear disappears; but, at the first light of dawn, you happen—to your great surprise—to gradually transform into a human being, while the strong pain you feel—caused by the metamorphosis—overrides your first surprise. When the Sun has fully risen your transformation is complete, and meanwhile the jungle is behind you, and you have reached a metropolis, within which you can see a large building, the city's university. You conceive that building as the place where you and other humans, as you now are, spend their time studying the universe, the universe which you belong to, of course. At that point, you can't avoid focusing on the mind-blowing relationship between your abstract thinking ability and that hard, solid stuff that makes up everything in the physical world around you, including the Sun in the sky, buildings on Earth, your own body, and your own ex-monkey brain, a brain of a sort of monkey that has lost its group. At that point, you understand that the whole situation, including the specific problem you are dealing with, is so absurd and marvellous at the same time, that it can only be a dream. So you actually wake up, and you understand that who you dreamt you were was not, in fact, as free to act as you felt they were, since the perceptions and thoughts of the monkey-human you dreamt of were just a portion of the whole dream that was taking place in your dreaming mind, and what you thought of as material in the dream was just the remaining portion of your dreaming mind, which was therefore, in fact, much bigger than you thought it was in the dream, and which mysteriously prepared unexpected and surprising scenarios for you that were only apparently physical. At this point, it may happen that you begin to seriously doubt that the comfortable bed you now lie on in the dark of the night, like everything else in the universe, is made of matter, and you begin to think that the condition you believe you are in now—you believe you are awake and free to act as you like—may be, all in all, not so different from some sort of dream happening in the mind of God.

Conclusions

The path we took in this work was basically the following. First of all, starting from the current ignorance with respect to the mind-matter problem, we explored the possibility that there is probably 'more mind' in the universe compared to what is currently thought. In practice, we have investigated the possibility that mind does not belong exclusively to certain biological systems, a possibility that has already been extensively investigated by philosophy of mind, since, with the end of vitalism, the brain no longer has anything special compared to other physical systems. What we have done, however, is to explore a typically unexplored region in philosophy of mind, that is, to look for mind in another 'natural system', i.e. in another system which is not a product of mind—as the brain is believed not to be—claiming that it is legitimate to do so at least as much as wondering if it is possible to design and build a thinking machine. As we identified this possible 'thinking system', we saw that we cannot be certain of the neo-Darwinian conclusions regarding the nature of the phenomenon of life and its origins, in the sense that at least the first living organisms on Earth could actually be artefacts, i.e. products of a mind. We have thus seen how this approach to the mind-matter problem is intertwined with the possibility of a deeper understanding of the phenomenon of life, which could in principle admit a semantic level of understanding. From this perspective, the phenomenon of life cannot be confined to what appeared on Earth 4 billion years ago, and plausibly on other planets. Rather, the phenomenon of life on Earth and possibly on other 'thinking planets' is seen as a reworking of a pre-existing phenomenon which involves the strong interaction and ultimately the quantumness of nature. Such a reworking, in which an 'electromagnetic version' of life is produced, takes place by virtue of a structure whose activity is basically of an electromagnetic nature, insofar as it can be triggered by particle showers, acting therefore as a strong-e.m. interface between the reality of a phenomenon and its elaboration and representation. As previously discussed, it is therefore possible to rewrite a typical abiogenetic history from a possible first-person perspective, i.e. the author's perspective. In this approach, unlike Nagel, we started from the assumption that there is nothing false or wrong in current neo-Darwinian science, fundamentally based on plausible abiogenic histories and on the mechanism of random mutation and natural selection. What we meant to question are the excessively hasty conclusions about the true nature of life, which typically overlook the fact that still nowadays we do not understand why the activity of a certain physical system — our brain at least — actually corresponds to a subjective experience of mental contents, which by the way are not objects of study of physico-chemical-biological science.

There is a risk that a residue of animism or paganism could be glimpsed in the proposed scenario, but to avoid this risk we have deconstructed in Chapter 2 the concept of authorship, precisely to emphasize that we do not deny that life on Earth appeared by virtue of nothing other than the mechanistic forces of nature.

Our position on the problem of the origins of life is that, if set in terms of the mind-matter problem, it may be more interesting than the way it is typically addressed in the theistic-atheistic debates, or science-faith debates, and that a similar approach is in principle possible when the debate concerns problems such as the fine-tuning of the fundamental constants of the universe.

The main risk incurred in addressing these arguments is certainly that of losing neutrality with respect to the conclusions; that is, it may happen that we tend more or less inadvertently to defend the vision of the world that somehow suits us best for personal reasons. As an example, it is certainly not easy to adopt a neutral and detached attitude towards the problem of the existence of our free will, or equivalently towards the possibility that we are indeed artificial intelligence machines, a drawback fortunately greatly reduced when the research concerns very technical and specific topics that do not concern us so closely. We do not hide, in this regard, the fact that one of the reasons that prompted us in the whole reflection was the need to produce a synthesis of the two main antithetical positions, namely meaningless materialist scientism on the one hand, and, on the other hand, any religious position that still tells a childish story about a supernatural being who creates freely acting humans and judges them for their conduct.

Regardless of the personal motivations which led to this work, along the path we have undertaken we have therefore observed that wherever the dividing line is drawn between 'conscious' and 'non-conscious' agents, we always find ourselves with the problem of explaining how matter can think. We have argued that the problem consists precisely in the way in which we conceptualize what exists independently of the mind of each conscious agent in terms of a hypothetical substance other than that of which the mental contents are made, called matter, which is where the mind-matter problem arises, historically, both in dualism and in materialism. For us, an acceptable possibility for avoiding the problem is some form of well-formulated ideal-realism, and we have only hinted at how this might be constructed, showing that, within such idealism, other problems that arise could actually be traced back to language, unlike in materialism.

We can't help observing that an ontology of this type is reflected in the writings of various mystics of most epochs and cultures, such as those taken up by Aldous Huxley in his essay "The Perennial Philosophy", where the leitmotif is clearly the identity of the Atman with the Brahman, translatable with the famous expression "You are that". Such an agreement with mystic literature of course is not reported as an argument in favour of the validity of such an ontology, as the value of introspective research in the search for truth is not so recognized today. In this regard, however, our position towards the mind-matter problem is that the scientific method based on empirical research and logical-deductive reasoning alone is not adequate to account for the problem. It is sufficient in this regard to consider that the first step of the scientific method, in investigating for example the nature of heat, is to quantify the temperature of a body by associating it with the length of a mercury column, therefore by cutting out our subjective bodily sensation of 'hot' at the very beginning of the research. Such a way of proceeding of course allows sensational successes within an object-based scheme, but completely leaves out subjective experience. We simply observe that the latter can only be directly investigated in the intrinsic condition of withdrawal and solitude, because of the privateness—or undetectability—of each of our minds. In support of our position with respect to the method, we observe that in the context of the contemporary approach to the mind-matter problem philosophers of the mind and neurophysiologists often start a blame game, where the former realize that philosophy of mind has reached such a level of complication that the author often struggles to understand their own writing about the nature of mental states and finally argues that it is up to neurobiology to deal with the problem, but obviously the latter deals with the study of the brain, which is a physical system, neither more nor less than a steam machine is, and, since steam machines—probably—do not think, neither should the brain, as currently conceived.

In other words, after having exposed the possibility that there is more mind in nature than what is currently conceived, we have concluded that the only way we can think of in order to avoid the mind-matter problem is to admit that there is nothing but mind, since it is quite inadmissible to say that there is nothing but matter, because of the epistemological primacy of mind. Materialism, as an object-based schema, is a wonderful model that has allowed enormous successes in reductionist terms, and, within it, it is possible to infer that a free-acting individual does not exist, while failing to explain anything that truly matters to us, such as our thoughts, emotions, desires and so on. It is a wonderful model for most purposes, just as Newtonian gravity is, but it is intrinsically highly problematic, since, in materialism, by definition, matter is real and ontological, while the experience of mind, of which we are certain, and which is the ultimate root of knowledge, is not, thus introducing an unbridgeable gap between the epistemic and the ontological levels. Regarding the aspects that can be personally difficult to face, it should be noted that in an ideal-realism of the type proposed, one faces not only the difficulty of accepting the absence of individual free will, but perhaps even more so, the struggle of accepting the dream-like nature of the so-called waking state. But, despite the difficulties, which may be due to deep-rooted beliefs, we have explored such an old ontology in the light of modern knowledge about the brain, since it seems to us that a certain coherence with what we are sure of can be discerned within it—alongside perhaps neutral monism—despite having devoted very little space to the ontological problem.

Beyond our brief proposal regarding a way to avoid the mind-matter problem, in this book we intended to emphasize that the mind-matter problem will have an impact on our conception of reality comparable to what the black-body spectrum problem had at the beginning of the last century: it was initially seen as a problem confined to a very specific field, but it had a strong impact on our whole conception of the world, leading to the appearance of the quantum conception of nature. We suggest that the problem of consciousness—like, of course, that of free will—is a problem of at least equal magnitude, because it will test the reference paradigm of the sciences themselves, which is now materialism. And it will probably not be just a revolution in a wide knowledge-field, but a change that will necessarily have repercussions on our way of life, on society's systems of punishment and reward. But, above all, it will lead to a human being at least psychically and culturally quite different compared to us: something like a physical person who is well aware of its impersonality—and maybe of its immateriality—and who has accepted it. In this regard, it should be noted that while contemporary physics differs substantially from Aristotelian-style folk physics, in which forces are proportional to speeds rather than to accelerations, in modern psychology this is in fact not the case at all, as the individual is still typically thought of as a free-acting individual, and in fact the effective reference paradigm of psychology is still Cartesian dualism, like that of folk psychology.

To conclude, the mind-matter problem is certainly interdisciplinary in itself, but it seemed to us that it was far broader than currently conceived, as we have attempted to show in this work. We admit that this is certainly an imperfect and incomplete work, which we cannot see as anything more than an opportunity to reflect upon what we actually know, compared to what we believe we know, in the manner of Socrates. However, in this regard, we can rejoice in being justified in believing that we don't have any real responsibility for our thought process; we just observed as bystanders what happened in our minds correspondingly to some blind, purposeless and meaningless electrochemical neural phenomenon, and just wrote about it as nothing but pure witnesses of the whole process.

Endnotes

Chapter 1

1 In eukaryotes it occurs across the inner mitochondrial membrane, while in prokaryotes, such as bacteria and archaea, it occurs across the cell membrane.

2 No known archaea carry out photosynthesis.

3 Most of the mass of matter (99%) is given by the strong interaction field energy; only 1% of the mass is given by the interaction of the fermion fields with the Higgs field, while the contribution of electromagnetic binding energy to the total mass of matter is of the order of 10^{-10}.

4 The verb 'to metabolize' is a Greek-derived form, while 'to transform' is derived from Latin.

5 We are omitting here the various types of external cell membrane and cell wall.

6 Of course there are many ways of defining what a gene is; this definition is fine for our purposes.

7 This gives rise to horizontal genetic transfers, precursor mechanisms of the vertical genetic transfer that occurs in the sexual reproduction of the most advanced organisms.

Chapter 2

1 Here the term "signal" refers to the way in which the term is used in physics and not in semiotics, in which it basically indicates an intentional sign.

2 Of course the easiest way to do it is to use a universal computer and to program it for the task.

3 Here we are forced to deal in an absolutely stringent way with an issue that of course has enormous consequences for our society; in fact, it pushes us to rethink the concept of moral responsibility and therefore to reflect, for example, on the merit-reward and guilt-punishment systems.

Chapter 3

1   The term "regions" from here onwards refers to thunderstorm convective cells. We preferred the term "regions" to avoid
· confusion with the cellular low-level structure of the system.

Chapter 4

1   Here we are referring to the tail of a shower induced by a high-energy cosmic ray, or equivalently to a little shower induced by a low-energy cosmic ray.

2   Even though this sounds like a finalistic description, we are not making any finalistic claims. The description is borrowed from biology just to show that it can be borrowed. Within biology itself, finalistic language is often used for convenience without actually sustaining any finalism.

3   The following expression is a rough evaluation because the physical beam is not monochromatic and the energy of every particle is also not constant in time. We are just interested in qualitatively seeing how exponential decay and growth concur in particle physics.

4   The total momentum conservation law is imposed and the electron mass is overlooked. The electron energy basically constitutes the energy present in the pion environment, of which only $2m_0c^2=270MeV$ is used for pion reproduction, i.e. is converted into pion mass.

5   Another peculiar characteristic of this object is that it does not need to absorb anything other than energy from its environment, while a cell must necessarily absorb matter too in order to reproduce itself.

6   In this regard, we mention the works by Frappat, Sciarrino and Sorba (Frappat, Sciarrino, and Sorba 2000; Sciarrino and Sorba 2014) in which a mathematical model is developed to explain the composition in degenerate multiplets of the genetic code in the context of the representation theory of symmetry groups along the lines of the derivation of hadron multiplets.

7 While avoiding defining a complexity metric, we only observe that anyone who has worked on a latex simulation of what happens in a neutral pion would certainly struggle to say that this is not a complex system.

8 The above authors also studied the stability of a string, which results in being stable at high temperatures.

Chapter 5

1 Asking the reason for the quantum nature of reality is like asking why the speed of light is constant in any reference frame, and, more generally, why the laws of physics are as they are. This problem is introduced as a purely literary expedient, to focus the attention on nature's inherent discreteness and, subsequently, on its consequences.

2 It should be noted that in nature different states of matter of the primordial expanding universe are more or less faithfully reproposed at hydrostatic equilibrium in stars by virtue of gravitational fields which provide counterpressure. The gravitational field is responsible for the hydrostatic equilibrium of the most common stars, in which conditions more or less similar to those of primordial nucleosynthesis are found, and of the neutron stars in whose nucleus QGP is believed to be present in conditions of lower temperature and higher chemical potential and pressure compared to primordial QGP. Here we are not interested in studying a physical object at the equilibrium that appears naturally in the universe, such as a star or a neutron star, but rather we are interested in studying from a theoretical point of view how a QGP fireball at the hydrostatic equilibrium would behave, postponing any questions on its actual feasibility.

3 This is a plausible imagined event that has the sole literary purpose of introducing the idea that we are going to expose.

4 The idea is obviously inspired by the role of ATP in living organisms, but we cannot explicitly refer to them in the text

because we are presenting a possible intention underlying the emergence of life, which therefore in this mental setting 'does not yet exist'.

5 We know that it can be done since in living organisms proteins are linear chains made up of 20 different amino acids which organisms synthesize more or less autonomously.

6 An effective implementation is of course the transcription-translation process described in fig. 1.4 in which data are first copied and then decoded since ribosomes (the decMachines) cannot directly read DNA (the data tape).

7 In biological terms, of course, in this scenario each molecular tape contains a gene.

8 Of course the above is probably too simple a version of a replisome, which requires kinases, elicases, polymerases, topoisomerases, primases, and so on. In the illustrated process, the replication of the two strands occurs in the same direction, in contrast to what happens in the living organisms that we know; this makes the replication mechanism simpler but less efficient because of the impossibility of introducing error-correction mechanisms during replication. Even the replication fork typically proceeds in opposite directions, as illustrated in Chapter 1. Furthermore, in living cells the kinases independently take care of transferring phosphate groups from ATP to the deoxyribonucleotides, without which the polymerization does not take place.

9 In other words, this machine operates as a Cre-recombinase in site-specific recombination, which is a basic mechanism of horizontal gene exchange.

10 Furthermore, the need for dedicated signalling molecules to move throughout the system also limits the size of the system.

11 That is why the extension of a cell typically ranges from 5 to 200 microns.

12 Such as that involved in plasmids partitioning mentioned in Chapter 1.

Chapter 6

1 Accordingly, the concentration of dNMPs is higher than in typical known organisms and this requires dedicated machinery. Otherwise, such a high concentration would present problems; e.g. nucleotides would chelate the magnesium necessary for replication.

Chapter 7

1 "For my part, when I enter most intimately into what I call myself, I always stumble on some particular perception, of heat or cold, of light or shadow, of love or hate, of pain or pleasure. I never catch myself without a perception, nor do I ever observe anything other than perceptions." David Hume, *A Treatise of Human Nature: Being an Attempt to Introduce the Experimental Method of Reasoning into Moral Subjects* (1739).
2 Interview in *The Observer* (25 January 1931), p. 17, column 3.
3 Certainly this problem does not appear in pluralistic idealism, but we believe that such a pluralistic ontology is too hard to sustain as of today.
4 "Physics and Reality" (1936), in *Ideas and Opinions*, trans. Sonja Bargmann (New York: Bonanza, 1954), p. 292.

Bibliography

Anderson, P. W., and D. L. Stein. 1987. "Broken Symmetry, Emergent Properties, Dissipative Structures, Life." In *Self-Organizing Systems*. Edited by F. E. Yates et al. Life Science Monographs. Boston: Springer, 445–57.

Arsene, I., I. G. Bearden, D. Beavis, C. Besliu, B. Budick et al. 2005. "Quark-Gluon Plasma and Color Glass Condensate at RHIC? The Perspective from the BRAHMS Experiment." *Nuclear Physics A* 757 (1-2 SPEC. ISS.): 1–27. https://doi.org/10.1016/j.nuclphysa.2005.02.130.

Balser, M., and C. A. Wagner. 1960. "Observations of Earth-Ionosphere Cavity Resonances." *Nature* 188: 638–41.

Bauer, Heidi, Heinrich Giebl, Regina Hitzenberger, Anne Kasper-Giebl, Georg Reischl, Franziska Zibuschka, and Hans Puxbaum. 2003. "Airborne Bacteria as Cloud Condensation Nuclei." *Journal of Geophysical Research: Atmospheres* 108 (D21). https://doi.org/10.1029/2003JD003545.

Baym, Gordon. 2016. "Ultrarelativistic Heavy Ion Collisions: The First Billion Seconds." *Nuclear Physics A* 956: 1–10. https://doi.org/10.1016/j.nuclphysa.2016.03.007.

Berera, Arjun, Robert Brandenberger, Joel Mabillard, and Rudnei O. Ramos. 2016. "Stability of the Pion String in a Thermal and Dense Medium." *Physical Review D* 94 (6): 065043. https://doi.org/10.1103/PhysRevD.94.065043.

Bodifée, G. 1986. "Star Formation Regions as Galactic Dissipative Structures." *Astrophysics and Space Science* 122 (May): 41–56. https://doi.org/10.1007/BF00654379.

Bystrai, G. P., I. A. Lykov, and S. A. Okhotnikov. 2011. "Thermodynamics of Nonequilibrium Processes in a Tornado: Synergistic Approach." *arXiv E-Prints*, September, arXiv:1109.5019. http://arxiv.org/abs/1109.5019.

Carlson, Neil N. 1996. "A Topological Defect Model of Superfluid Vortices." *Physica D: Nonlinear Phenomena* 98 (1): 183–200. https://doi.org/10.1016/0167-2789(96)00052-8.

Crick, F., and C. Koch. 1998. "Consciousness and Neuroscience." *Cerebral Cortex* 2 (8): 97–107.

Dawkins, Richard. 2017. *The Blind Watchmaker*. Milan: Mondadori.

Dennis, A. S. 1970. "The Flashing Behavior of Thunderstorms." *Journal of Atmospheric Sciences* 27 (1): 170–2. https://doi.org/10.1175/1520-0469(1970)027<0170:TFBOT>2.0.CO;2.

Dwyer, J. R., and M. A. Uman. 2014. "The Physics of Lightning." *Physics Report* 534 (January): 147–241. https://doi.org/10.1016/j.physrep.2013.09.004.

Franzblau, Edward, and Carl J. Popp. 1989. "Nitrogen Oxides Produced from Lightning." *Journal of Geophysical Research: Atmospheres* 94 (D8): 11089–104. https://doi.org/10.1029/JD094iD08p11089.

Frappat, Luc, Antonino Sciarrino, and Paul Sorba. 2000. "Crystal Basis Model of the Genetic Code: Structure and Consequences." *Proceedings of Institute of Mathematics of NAS of Ukraine*, 481–8.

Heinz, Ulrich, and Maurice Jacob. 2000. "Evidence for a New State of Matter: An Assessment of the Results from the CERN Lead Beam Programme." https://arxiv.org/abs/nucl-th/0002042.

Huang, Tao, Yunde Li, Hong Mao, Michiyasu Nagasawa, and Xinmin Zhang. 2005. "Signal of the Pion String at High-Energy Collisions." *Physical Review C* 71 (1): 014902. https://doi.org/10.1103/PhysRevC.71.014902.

Hutto, Daniel. 1998. "An Ideal Solution to the Problems of Consciousness." *Journal of Consciousness Studies* 5 (January): 328–43.

Kandel, Eric R., James H. Schwartz, and Thomas M. Jessell, eds. 1991. *Principles of Neural Science*. Third Edition. New York: Elsevier.

Karouby, Johanna. 2013. "String Melting in a Photon Bath." *Journal of Cosmology and Astroparticle Physics* 2013 (10): 017. http://stacks.iop.org/1475-7516/2013/i=10/a=017.

Kim, Jaegwon. 1995. "Mental Causation in Searle's 'Biological Naturalism'." *Philosophy and Phenomenological Research* 55 (1): 189–94. http://www.jstor.org/stable/2108318.

Kondepudi, Dilip, and Ilya Prigogine. 2014. *Modern Thermodynamics: From Heat Engines to Dissipative Structures.* Chichester/New York: John Wiley & Sons.

König, H. L., and F. Ankermüller. 1960. "Über den Einfluss besonders niederfrequenter elektrischer Vorgänge in der Atmosphäre auf den Menschen." *Naturwissenschaften* 21: 486–90.

König, H. L., S. Lang, and A. P. Krueger. 1981. *Biologic Effects of Environmental Electromagnetism.* New York: Springer-Verlag.

LaHurd, Danielle, and Corbin E. Covault. 2018. "Exploring Potential Signatures of QGP in UHECR Ground Profiles." *Journal of Cosmology and Astroparticle Physics* 2018 (11): 007. https://doi.org/10.1088/1475-7516/2018/11/007.

Libet, Benjamin, Curtis A. Gleason, Elwood W. Wright, and Dennis K. Pearl. 1983. "Time of Conscious Intention to Act in Relation to Onset of Cerebral Activity (Readiness Potential): The Unconscious Initiation of a Freely Voluntary Act." *Brain* 106 (3): 623–42. https://doi.org/10.1093/brain/106.3.623.

Martinez, Gines. 2013. "Advances in Quark Gluon Plasma." *arXiv E-Prints*, April, arXiv:1304.1452. http://arxiv.org/abs/1304.1452.

Mazur, Vladislav. 1982. "Associated Lightning Discharges." *Geophysical Research Letters* 9 (November): 1227–30. https://doi.org/10.1029/GL009i011p01227.

Milikh, Gennady, and Robert Roussel-Dupré. 2010. "Runaway Breakdown and Electrical Discharges in Thunderstorms." *Journal of Geophysical Research* 115 (December). https://doi.org/10.1029/2009JA014818.

Monod, Jacques. 1971. *Chance and Necessity*. New York: Vintage Books.

Nagasawa, M., and R. Brandenberger. 1999. "Stabilization of Embedded Defects by Plasma Effects." *Physics Letters B* 467 (3): 205–10. https://doi.org/10.1016/S0370-2693(99)01140-5.

Nagel, Thomas. 1974. "What Is It Like to Be a Bat?" *Philosophical Review* 83 (October): 435–50. https://doi.org/10.2307/2183914.

Nagel, Thomas. 2012. *Mind and Cosmos: Why the Materialist Neo-Darwinian Conception of Nature Is Almost Certainly False*. New York: Oxford University Press. https://books.google.it/books?id=pOzNcdmhjIYC.

Newman, M. E. J., S. H. Strogatz, and D. J. Watts. 2001. "Random Graphs with Arbitrary Degree Distributions and Their Applications." *Physical Review E* 64 (2). https://doi.org/10.1103/physreve.64.026118.

Persinger, Michael. 2012. "Brain Electromagnetic Activity and Lightning: Potentially Congruent Scale-Invariant Quantitative Properties." *Frontiers in Integrative Neuroscience* 6: 19. https://doi.org/10.3389/fnint.2012.00019.

Pobachenko, S. V., A. G. Kolesnik, A. S. Borodin, and V. V. Kalyuzhin. 2006. "The Contingency of Parameters of Human Encephalograms and Schumann Resonance Electromagnetic Fields Revealed in Monitoring Studies." *Journal of Biophysics* 51 (3): 480–3.

Prigogine, Ilya, and Isabelle Stengers. 1984. *Order Out of Chaos: Man's New Dialogue with Nature*. New York: Bantam Books.

Raymond, Jason, Janet L. Siefert, Christopher R. Staples, and Robert E. Blankenship. 2004. "The Natural History of Nitrogen Fixation." *Molecular Biology and Evolution* 21 (3): 541–54. https://doi.org/10.1093/molbev/msh047.

Saroka, K. S., and M. A. Persinger. n.d. "Quantitative Evidence for Direct Effects between Ionosphere Schumann Resonances and Human Cerebral Cortical Activity." https://www.researchgate.net/publication/277934441_Quantitative_

Evidence_for_Direct_Effects_between_Earth-Ionosphere_ Schumann_Resonances_and_Human_Cerebral_Cortical_ Activity.

Schrödinger, Erwin. 1951. *My View of the World*. Cambridge: Cambridge University Press. https://doi.org/10.1017/ CBO9781107049710.

Schumann, W. O. 1952. "Über die strahlungslosen Eigenschwingungen einer leitenden Kugel, die von einer Luftschicht und einer Ionosphärenhülle umgeben ist." *Zeitschrift für Naturforschung A* 7 (2): 149–54.

Schumann, W. O., and H. König. 1954. "Über die Beobachtung von 'Atmospherics' bei geringsten Frequenzen." *Naturwissenschaften* 8: 183–4.

Sciarrino, Antonino, and Paul Sorba. 2014. "Crystal Basis Model: Codon-Anticodon Interaction and Genetic Code Evolution." *P-Adic Numbers, Ultrametric Analysis, and Applications* 6 (June): 257–74. https://doi.org/10.1134/S2070046614040013.

Searle, John R. 1992. *The Rediscovery of the Mind*. Cambridge, MA: MIT Press.

Smyth, William D., and James N. Moum. 2012. "Ocean Mixing by Kelvin-Helmholtz Instability." *Oceanography* 25 (2): 140–9. https://doi.org/10.5670/oceanog.2012.49.

Stock, Reinhard. n.d. "7 Relativistic Nucleus-Nucleus Collisions and the QCD Matter Phase Diagram." Datasheet from Landolt-Börnstein – Group I 'Elementary Particles, Nuclei and Atoms', Volume 21A: *Theory and Experiments*. Springer Materials. Edited by H. Schopper. Berlin/Heidelberg: Springer-Verlag. https://doi.org/10.1007/978-3-540-74203-6_7.

Taylor, J. R. 1972. *Scattering Theory: The Quantum Theory on Nonrelativistic Collisions*. New York: Wiley.

Trautteur, Giuseppe. 2009. "The Illusion of Free Will and Its Acceptance." In *After Cognitivism*. Edited by Karl Leidlmair. Dordrecht: Springer, 191–203. https://doi.org/10.1007/978-1-4020-9992-2_12.

Trautteur, Giuseppe. 2020. *Il Prigioniero Libero*. Milan: Adelphi.

Vachaspati, T. 1992. "Vortex Solutions in the Weinberg-Salam Model." *Physical Review Letters* 68 (13): 1977–80.

Vilenkin, A., and E. P. S. Shellard. 2000. *Cosmic Strings and Other Topological Defects*. Cambridge: Cambridge University Press. http://www.cambridge.org/mw/academic/subjects/physics/theoretical-physics-and-mathematical-physics/cosmic-strings-and-other-topological-defects?format=PB.

Vonnegut, B., O. H. Vaughan Jr, M. Brook, and P. Krehbiel. 1985. "Mesoscale Observations of Lightning from Space Shuttle." *Bulletin of the American Meteorological Society* 66 (1): 20–9. https://doi.org/10.1175/1520-0477(1985)066<0020:MOOLFS>2.0.CO;2.

Wilczek, Frank. 2016. *A Beautiful Question: Finding Nature's Deep Design*. New York: Penguin Random House.

Yair, Yoav, Reuven Aviv, Gilad Ravid, Roy Yaniv, Baruch Ziv, and Colin Price. 2006. "Evidence for Synchronicity of Lightning Activity in Networks of Spatially Remote Thunderstorms." *Journal of Atmospheric and Solar-Terrestrial Physics* 68 (12): 1401–15. https://doi.org/10.1016/j.jastp.2006.05.012.

Yair, Yoav, Peter Israelevich, Adam D. Devir, Meir Moalem, Colin Price, Joachim H. Joseph, Zev Levin, Baruch Ziv, Abraham Sternlieb, and Amit Teller. 2004. "New Observations of Sprites from the Space Shuttle." *Journal of Geophysical Research (Atmospheres)* 109 (D15): D15201. https://doi.org/10.1029/2003JD004497.

Yair, Yoav Y., Reuven Aviv, and Gilad Ravid. 2009. "Clustering and Synchronization of Lightning Flashes in Adjacent Thunderstorm Cells from Lightning Location Networks Data." *Journal of Geophysical Research: Atmospheres* 114 (D9). https://doi.org/10.1029/2008JD010738.

Zhang, Xinmin, Tao Huang, and Robert H. Brandenberger. 1998. "Pion and η' Strings." *Physical Review D* 58 (2): 027702. https://doi.org/10.1103/PhysRevD.58.027702.

ESSENTIA

Essentia Books, a collaboration between Collective Ink and
Essentia Foundation, publishes rigorous scholarly work relevant to
metaphysical idealism, the notion that reality is essentially mental
in nature. For more information on modern idealism, please visit
www.essentiafoundation.org.